D0667043
DATE DUE

John Tirman is senior editor and head of the communications group of the Union of Concerned Scientists, Cambridge, Massachusetts. Tirman holds a Ph.D. in political science from Boston University, where he was also a Teaching Fellow of the College of Liberal Arts, 1972-75. He has worked as a political consultant, a reporter for *Time* magazine, and as senior policy analyst for the New England Regional Commission. His articles on energy, arms control, and science policy have appeared in *Technology Review,* the *New York Times,* the *Washington Post, The Nation,* and many other magazines.

THE MILITARIZATION OF HIGH TECHNOLOGY

THE MILITARIZATION OF HIGH TECHNOLOGY

Edited by
JOHN TIRMAN

BALLINGER PUBLISHING COMPANY
Cambridge, Massachusetts
A Subsidiary of Harper & Row, Publishers, Inc.

Chapter 2 is adapted from "Hi-Tech Warfare," which appeared in *The New Republic*, November 1, 1982. It is reprinted by permission of the author and The New Republic, Inc.

Chapter 3 is a revised version of a chapter appearing in the Club of Rome report, *Microelectronics and Society*, and is reprinted by permission of the author and Pergamon Press, Ltd., Oxford, England.

Chapter 5 is adapted from the Council on Economic Priorities report, *Military Expansion, Economic Decline.* It has been revised by the author and is reprinted by permission of the Council on Economic Priorities, 84 Fifth Avenue, New York, N.Y. 10014.

Chapter 9 is adapted from the Highlander Center report, *Our Own Worst Enemy: The Impact of Military Production on the Upper South*, and is published by permission of the author.

"The Darpa Exception," pp. 222–225, is adapted from the authors' forthcoming book, *The Innovators: Rediscovering America's Creative Energy* (Harper & Row), and is published by permission of the authors.

International Standard Book Number: 0-88410-947-X

Library of Congress Catalog Card Number: 84-12404

Printed in the United States of America

Library of Congress Cataloging in Publication Data

Main entry under title:

The Militarization of high technology.

Includes index.
1. Munitions—United States. 2. High technology industries—United States. I. Tirman, John.
HD9743.U6M53 1984 338.4'36213817'0973 84-12404
ISBN 0-88410-947-X

CONTENTS

LIST OF TABLES

PREFACE

Soon after Ronald Reagan assumed the presidency in January 1981, it was apparent that he intended to carry out his campaign promise of a substantial "rearmament"—a major increase in military spending to build up America's conventional and nuclear arsenals. The initial estimates showed Department of Defense budgets rising by $181 billion more than the Carter administration's projections in the 1982 to 1986 fiscal years, a total of some $1.5 trillion. By the time the fiscal year 1985 budget was being proposed, the figure had risen again: for the years 1984 to 1988, Defense Department budgets were slated to exceed $1.8 trillion. And these awesome figures did not include military-related spending in other executive departments. Clearly, a military program of massive proportions was shaping up in the Reagan White House.

The size and scope of the Reagan defense budgets quickly sparked a hot debate: Could the economy "absorb" such expenditures? What would the effects of these large Pentagon outlays have on employment, inflation, the federal deficit, and interest rates? Was American military preparedness and U.S. foreign policy in such a shambles that this extraordinary cash rescue was needed?

The dangers inherent in such a buildup seemed apparent. Conventional wisdom held that the recurrent price inflation experienced in the 1970s was the result of the precipitous rise in military spending during the Vietnam War. President Johnson was unwilling to raise taxes to pay for the Indochina conflict, and the economic consequences were harsh. In fact, the Vietnam buildup cost only one-third of the Reagan rearmament in constant dollars, comparing the 1965–70 period with 1981–86, so the concern voiced by many economists in the early 1980s seemed well-founded. And new conditions in the economy reinforced that worry: when Reagan took office, interest rates and federal budget deficits were already at record high levels.

A more specific kind of danger loomed as well. The potential "macro" effects of defense spending were troubling enough, but the nature of the rearmament was strongly oriented toward the procurement of new weapons and other capital goods, instead of toward the other large shares of the defense budget—operations and maintenance, and personnel. Procurement, which nearly doubled from 1981 to 1984, has a special meaning for the economy. The products being procured are technologies, often very sophisticated technologies, to be used in weapons, communications, transportation, and many other applications. In particular, the Pentagon would be seeking "high technology"—microelectronics, precision instruments, avionics, and the like—which is being integrated into military systems with great enthusiasm.

The prospect of a growing Pentagon–high-tech partnership is unsettling to many economists and technologists, because high tech is now regarded as being one of America's strong suits in world trade. The Defense Department's thirst for scientists and engineers, plant capacity, innovation, and research and development funds is viewed as being in direct competition with civilian, or commercial, high-tech applications. The high-tech industries, if stretched to capacity (in skilled employees, investment, production facilities) to compete in commercial markets, may be seriously strained when the Pentagon enters with its tens of billions of dollars in orders, its urgent timetables, and its exacting technical specifications. If a firm in Minneapolis has a defense contract to produce an unusual infrared imaging device for satellite reconnaissance, it may *not* be producing something for the overall commercial market—an imaging technology for medical applications, perhaps. Critics contend that such a tradeoff has a multiplicity of bad side-effects. The danger in this for America is that Japan or Germany (or some other country) may produce the medical imaging machine instead, bolstering its position in world trade; the danger to the firm is that when the Pentagon pulls the plug on its contract, it may be left without the necessary skills or capital to shift to commercial products; the danger to the scientists and engineers in the firm is that they may become too specialized, attuned to the rigorous demands of military R&D, and their shift to commercial work may be impaired as well. And so on. The result: American high-tech industries in turmoil, behind schedules, drawn to military production rather than commercial innovation, and increasingly at a disadvantage compared with the mounting challenge from Japan and other global competitors.

This is, to put it mildly, a controversial viewpoint. It is only within the last decade or so that military spending has been considered as potentially harmful to the economy, rather than as a benefit. As the growth of the U.S. economy stalled repeatedly in the 1970s and 1980s, the effects of the Vietnam buildup on inflation were frequently cited. The idea that the employment effect of a dollar spent on defense was less socially useful than a dollar spent on, say, health care also gained currency. More often, however, the debate took the form of "social needs" versus the military. The growing share of federal budgets dedicated to the Pentagon was viewed by liberals as being an encroachment on social programs—for example, income transfers, Social Security, nutrition subsidies—and federal commitments to education, transportation, energy, housing, basic scientific research, and a myriad of other nonmilitary priorities. The perceived tradeoff between social and military expenditures became particularly polarized in the Reagan administration, due to its stated intent to reduce growth in social programs (and, in some cases, to reduce real-dollar outlays to social programs, or to eliminate a few such projects outright) while increasing military strength correspondingly. Indeed, this hot debate pitting human needs against the demands of the military establishment dominated discussion about the Pentagon's growing budgets. But the notion that military spending—and procurement in particular—had inherently inimical effects on the economy, and would hurt technological development, is less widely held. Indeed, if anything, the military's involvement with technological innovation was usually looked at in a positive light—commercial products were often cited as "spinoffs" from Pentagon projects, for example, or the survival of whole industries (like aviation) was attributed to their ties to defense.

This book will examine the relationship between the U.S. military and American high technology industries. In Chapter 1, I will review and assess the general debate over the military's effects on the economy and technology, how that debate has evolved over the last twenty years, where it stands now, and the direction it seems to be headed. Chapters 2 and 3 will also serve as overviews. Robert Reich briefly contrasts military and trade policy, showing how the Reagan administration's emphasis on the former may subvert the chances of success in the latter. The essay was written in the midst of the controversy over European sales of pipeline technology to the Soviets, which still serves as a sharp example of the American dilemma. More important, Reich's chapter addresses the general problem that the stress created by military demands on high tech may place commercial high tech at a disadvantage in global markets. The third chapter, written by Frank Barnaby, reviews the uses of high technologies—particularly microelectronics—in military hardware, showing the extent of the Pentagon's employment of advanced technology, how that use appears to be accelerating, and the ways in which a high-tech military can affect strategic doctrine and the prospects for war.

Chapters 4 and 5 focus on the "spinoff" argument. Gordon Thompson demonstrates how the military played a decisive—and perhaps deleterious—role in the genesis of an advanced technology of thirty years ago: nuclear power. Robert DeGrasse evaluates the influence of the Defense Department in the development of semiconductors. Both of these historical cases show that the Pentagon's involvement in technological evolution—and the benefits or drawbacks accruing from that involvement—cannot be understood in black and white terms. Assessing the influence of the military on technology is difficult and complex, even with the benefit of hindsight.

Our analysis then turns to the individual institutions and people affected by the Pentagon-high-tech partnership. John Ullmann looks at the problems surrounding civilian firms involved in military work, and Lloyd Dumas examines the effect of military contracting in America's research centers—university and industrial—in order to assess the Pentagon's impact on innovation. In Chapter 8, Warren Davis explores the role of the scientist in defense work, the specific ways a scientist may become ensnared in the industry, and, importantly, the moral dimension affecting the individual. Tom Schlesinger then discusses the increasing emphasis on automation in defense production and its impact on labor and regional development.

In a final chapter, I will draw the analyses together to suggest a pattern to the Defense Department's impact on advanced technology, with special reference to a developing high-tech industry—biotechnology. In addition, Chapter 10 will include a special section by James Botkin and Dan Dimancescu that discusses an alternative model for military procurement.

The bias of the book will be quickly apparent. It is not a bias "against" military spending per se, or against the use of high technology in military technology, or against adequate military strength. These issues are important, and will be mentioned from time to time—one can hardly avoid confronting such basic policy issues—but the purpose of the book is not to answer these questions as such. Rather, it is to argue this somewhat new notion that military procurement has a profound, perhaps damaging, effect on the vitality of commercial high technology. The extent to which sophisticated technologies *should* be integrated with weaponry, communications, and the like is not a topic of this collection. It is merely shown—especially in Mr. Barnaby's essay—that high tech is now at the very core of military planning and weapons development. Nor do we evaluate whether or not the levels of military spending for procurement are excessive; that would demand a very different sort of book, one that analyzed exhaustively the nature of American security needs. The contributors to this volume, however, do imply that the current rearmament, one likely to continue throughout the 1980s regardless of who is in the White House, is excessive. We believe that even if one does agree that U.S. security needs bolstering, such strengthening of American security could be accomplished in a way that does not jeopardize the nation's economic competitiveness at the same time.

Finally, it must be stated from the outset that this question of the military's impact on technology is not one readily answered by empirical evidence. Although every effort has been made by the contributors to employ available data that bear on each argument, the subject is almost too large to be "provable" either way. One cannot say with complete confidence that the military's impact on, say, the history of aviation has been positive, because we don't know what would have happened to aviation if the military had not played such a significant part. In the same way, we cannot predict that a particular researcher would produce something commercially worthwhile in the absence of military support. Such are always the unknowables in this kind of analysis. Yet the difficulty of this book's assessment is compounded by military secrecy, the very scope of the military enterprise, the inadequacy of economic data collection and analysis, the rapidly changing structure of high technology industries, and the relationship of the Department of Defense to those industries.

As a result, the "answer" we collectively derive and share is a somewhat tenuous and largely suggestive one. But we maintain that the present mode of procurement, and the very size and nature of this procurement, do have a harmful effect on commercial high technology, on the individual firms and people employed to produce military high tech, and on America's economic position in the world.

* * *

I wish briefly to thank the several people involved in this modest project. In addition to the contributors, thanks are due to Carol Franco, my editor at Ballinger, and her colleagues; David Cohen, an editor at *The Boston Globe Magazine*, who worked with me on an article on the same subject that sparked Ms. Franco's interest; and to my co-workers at the Union of Concerned Scientists, a constant source of stimulation and information.

Thanks of a different order are sent to Carol Flake, and to my mother and father.

It is my hope that this book answers some questions, poses others, but, most important, spurs the reader to carry the questions and answers on to more study and action.

John Tirman
Cambridge, Massachusetts
February 1, 1984

1 THE DEFENSE-ECONOMY DEBATE

John Tirman

INTRODUCTION

> Because the dollars spent on defense go for the production of material—thus creating jobs, stimulating investment, and generating individual and corporate tax revenues—cutting defense spending is an inefficient way to decrease the deficit. On the other hand, transfer payments have a direct effect on the deficit. They rarely provide tax revenues. And they usually are spent on non-durable goods, with a lower investment component. Thus they have a lower multiplier effect than defense spending.
>
> Transfer payments are usually those identified with social programs, and there are those who charge that defense spending is crowding out social spending. Their arguments are rooted in emotion, not fact.[1]

The words are those of Defense Secretary Caspar Weinberger in an address delivered to a group of Miami businessmen in September 1982. In that speech, and many more like it by the Pentagon's civilian chief in the Reagan administration's early years, Weinberger engaged in one of the most virulent debates in contemporary America—the impact of military spending on the nation's economic health and well-being. In the above two paragraphs he touched upon several of the debate's salient issues: the competing uses of government revenues, the ways in which particular kinds of spending affect the economy, the Pentagon budget's impact on federal deficits, and the supposed tradeoff between government programs. Secretary Weinberger's brief depiction of the military spending debate in the early 1980s was prejudiced, of course, but he did phrase the debate as it was commonly understood.

The principal worry of economists and a growing number of members of Congress was that the sharp rise in military spending would enlarge federal deficits to record levels. This, in turn, would mean that the U.S. Treasury would have to borrow more money to pay its bills, and that public borrowing would compete with private borrowing. Such competition for capital, it was feared, would drive up interest rates. High interest rates could increase the cost of doing business and choke off an economic recovery from the recession that was underway soon after Mr. Reagan took office. Large federal deficits could also create the notorious "chaser effect": by pumping billions of dollars into the economy via government spending (of whatever kind) without a concurrent rise in production, there would be too many dollars chasing too few goods, thereby bidding up prices for a broad range of consumer products and services. Result: the "stagflation" that beleaguered the U.S. economy through much of the 1970s—inflation with tepid, or negative, economic growth. The administration did not argue against the possible effects of large deficits—Mr. Reagan had, after all, campaigned on the theme of excessive federal deficits for a number of years—but pointed instead to "social" spending as the culprit. More important, while the president's men acknowledged that the Reagan "rearmament" was ambitious, and would require a certain measure of sacrifice in other budget areas, they contended the military buildup and the money needed to pay for it were imperative. National security could not be gauged in dollars and cents. The general consensus—if there was one—was perhaps best stated by the Congressional Budget Office in its 1983 study:

> The ultimate decision on procurement and other defense spending should depend on considerations of national security and priorities for the use of resources. Current forecasts suggest that the proposed rapid defense buildup need not rekindle inflation in the near term. The buildup could, nonetheless, contribute to tightness in some particular industries that do a great deal of defense work. This could raise risks of cost growth and delivery delays in weapon systems. Moreover, a defense buildup financed by large federal deficits that continue even after the economy recovers could damage economic performance in the long run.[2]

The debate of the early 1980s was tinged by America's relative decline as the world's dominant economy. For a number of reasons, the United States had not kept pace over the previous decade with Japan and Germany in productivity increases, trade, and overall prosperity. Internally, the searing price inflation that had beseiged the Nixon, Ford, and Carter administrations had resulted in a zero net increase in consumer buying power during the 1970s, after the nation had enjoyed rapid and sustained increases throughout the previous two decades. So the character of the defense-economy debate had assumed a fresh, and mean-spirited, character by the time Ronald Reagan was elected president. Unlike Lyndon Johnson's hope of having guns and butter, few assumed that both were possible in the new atmosphere of public austerity. It seemed that in a time of economic turmoil and a broad mandate to limit government spending, something had to go, either the guns or the butter. The Reagan team chose to keep the guns, and the ensuing debate inevitably revolved around that choice.

That tradeoff between social and military priorities, however, was something of a red herring. Not that the tradeoff was not real—several analysts pointed out that the decline in social program funding was nearly identical to military spending rises—but that the defense-economy debate had traditionally entertained a different set of problems and prospects, namely, that defense spending (regardless of its relationship to other federal programs) has an *intrinsic* effect on the economy. Moreover, the debate looked to specific forms of defense spending for specific effects. The ways in which the Pentagon deploys its dollars will, for example, influence the labor market, regional development, price inflation, the demand for capital goods, the course of technological development, and even America's place in world trade.

These kinds of effects took on more prominence as the administration's defenders built a persuasive case that the rise in the military budgets was not likely to be catastrophic if one considered the military budgets narrowly in relation to total national output. "All of this fear about what is going to happen to the American economy if spending increases by another 1 or 2 percent of G.N.P. is absolutely ridiculous," said economist Herbert Stein, once a top adviser to President Nixon. "The difference will hardly be noticed."[3]

Such optimism belied a tangible queasiness among economists of all political stripes about the procurement plans of the Department of Defense (DoD). For the rises in DoD spending were aimed to acquire weapons, communications equipment, and the like—capital goods to be produced largely in the nation's private sector. Even President Reagan's chairman of the Council of Economic Advisers, Murray Weidenbaum, voiced such a fear shortly after leaving his Washington post in 1983: "It is clear that the downturn in the economy has freed up excess capacity for defense," he told a congressional committee, "but I'm concerned that as the defense program really hits stride there won't be enough capacity, and that it will push up prices." Indeed, the *only* negative effect of the accelerated DoD spending program acknowledged officially was this "bottleneck" problem.[4]

Again, however, this was considered to be merely the price of neglect; the administration noted repeatedly that the rearmament was necessary because America had ignored its defense procurement needs in the post-Vietnam era, and that the president was merely trying to catch up. As Secretary Weinberger asks, "Even if achieving the security needed for [peace with freedom] meant some dislocations in the economy—and it does not today—who would suggest any other course?"[5]

What the bottleneck argument brought to the fore, however, was the very notion that specific kinds of defense spending have particular kinds of effects. In accounting for the bottlenecks, moreover, the administration also underscored the importance of defense planning. For if one subtracts from the federal budget the so-called self-funding income transfer programs, like Social Security, DoD represents 43 percent of federal spending, spending that is not merely transfers of money, but targeted, production-oriented spending—in short, an economic planning system.[6] These two elements lay at the heart of the defense-economy debate. First, to understand the most profound impact of defense spending, we must look to particular activities within the American economy such as technological development and not merely the "macro" indicators of unemployment, inflation, and the like. Second, the Department of Defense is actually a planning system, indeed, an immense planning system that is larger than any other single economic entity in the noncommunist world.

A BRIEF HISTORY OF THE DEBATE

The controversy over how military spending affects the American economy is largely a post–World War II debate, and for good reason. It has only been since America's entrance into that war that the United States has maintained the status of a world military power. It is of more than incidental interest, too, that the United States was in the midst of the Great Depression when war broke out in Europe in 1939. Although President Franklin Roosevelt's intervention to revive the sagging economy via the New Deal had produced some salutary effects, it was widely acknowledged that war production was responsible for pulling the nation out of the Depression. And in the aftermath of the war it was widely expected that America would return to economic hardship, as it did: during the general demobilization of the late 1940s, a severe recession settled over the U.S. economy, though it did not approach the depths of the Great Depression.

With the outbreak of new hostilities in Greece, Indochina, and, most important, Korea, America again mobilized militarily, and, from the second Truman administration on, never demobilized as it had in 1946–47. Whatever the causes and justifications of America's involvement around the world, the simple fact remained that the U.S. military had assumed a new prominence internationally. And with that rôle came costs, high costs for readiness, new weapons, and worldwide deployment of troops. A high level of military spending became a permanent fixture in the American economic landscape. And an unprecedented level of prosperity seemed to accompany this new role as the world's policeman.

Since the mid-1930s, the theories of John Maynard Keynes were gaining currency among American economists, theories that seemed well-suited to explain the phenomenon of military spending coupled with strong economic growth. The quagmire of the Depression had been the result of too little consumer demand: the collapse in the capital markets and the high rate of business failures in 1929–32 had created an enormous idle production capacity and 25 percent unemployment. The last factor meant that consumer demand was unable to pull the entire system to its feet. Without demand, factories would remain closed, more businesses would fail, and so on, in an endless cycle. Keynes's answer was for government to create the demand by pumping money into the economy. It could do so by transfer payments (Social Security, unemployment compensation), or by government projects of various kinds that would both pay wages and create a demand for capital goods.

The formula was only a minor success during the 1930s; its effect was limited in part because the government cash rescue was haphazard and relatively small. After Pearl Harbor, government spending was considerably greater, and the economy was suddenly operating at full throttle, a pace sustained throughout the 1941–45 American involvement in the war. After Hiroshima and the gradual but pervasive demobilization in Europe and the Pacific, the slack in the economy again appeared, bringing with it unemployment and declining economic activity. Only with the advent of the Cold War and the Korean conflict did America's economic prospects brighten. Once a permanent state of preparedness was established in the 1950s, an apparent state of prosperity was established with it. A sort of unintentional Keynesianism had rescued the economy, a variant of Keynes in which the large demand-creating government intervention into the economy was almost wholly military in origin.

It was in this seemingly unambiguous context that the first salvos in the defense-economy debate were tossed. Military spending had clearly rescued the economy from perhaps a perpetual state of crisis, depression, tragic levels of unemployment—all the vicissitudes associated with the dynamic of capitalism. A commonly held belief during the Great Depression was that the end of capitalism was near, that the collapse in the system long predicted by Marxists was underway. World War II was the first hiatus in that historic collapse; the Pax Americana following the war was the second. Indeed, the leftist critics saw the new world role of American military might as a twofold strategy of imperialism abroad and a bail-out for capitalism at home. Two formodable Marxist economists articulated this bold view: Paul M. Sweezy and Paul A. Baran.

Sweezy was a well-known scholar, a member of the Harvard faculty in the 1930s and 1940s, and the editor of the *Monthly Review*, a journal of Marxian economic analysis. An economics professor at Stanford, Baran was the more pro-Soviet of the two men, and, in many estimates, the more brilliant. Their collaboration in the late 1950s and early 1960s produced one of the seminal volumes of Marxian thought in post-war America, *Monopoly Capital: An Essay on the American Economic and Social Order*, which was first published in 1966, two years after Baran's death. The book is a terse assessment of U.S. politics and economic activity. It purports to explain, in Marx's language, the nature of "surplus value"—the difference between the labor value and materials of a commodity and its price, what the revisionists Baran and Sweezy called "economic surplus"—in advanced capitalism. In this view, it was the underconsumption of economic surplus that characterized the Great Depression: capital was not invested, products were not purchased, and, as a result, labor was not employed. What separated prewar, Depression-laden America from postwar prosperity was the military spending attending the new, world "imperialist" role assumed by the United States. "In 1939," they wrote, "17.2 percent of the labor force was unemployed and about 1.4 percent of the remainder may be presumed to have been employed producing goods for the military. . . . In 1961 (like 1939, a year of recovery from a cyclical recession), the comparable figures were 6.7 percent unemployed and 9.4 percent dependent on military spending."[7] The inescapable conclusion was that the new Department of Defense was expending, or absorbing, the economic surplus to ensure an adequate level of consumption. The American "oligarchy"—what Marx called the bourgeoisie—would not allow large civilian government spending for housing, food, health care, and the like, because those would compete with private enterprise and might compromise its class interests. Military spending was different: "It is obvious that the building up of a gigantic military establishment neither creates nor involves competition with private enterprise," the two economists contended. Moreover, "there is no doubt that supplying the military is universally regarded as good business: all corporations, big and little, bid for as large a share as they can get. The private interests of the oligarchy, far from generating opposition to military spending, encourage its continuous expansion."[8]

The weight of the cultural and political associations of patriotism and militarism could be brought to bear on this government imperative as well; whereas social spending could be denigrated as "handouts" to the slovenly, military spending was cast in the mold of national security and American global power. It was thus irresistible, and, in Baran and Sweezy's view, the military's expansion served the elite's economic interests at home and its widening interests abroad.[9] The military's large budget, sustained at a relatively constant plateau since the late 1940s, could not of itself maintain prosperity, however. The capital-intensive nature of military preparedness in the 1960s led Baran and Sweezy to observe that defense could not employ as many people as it did even a decade before, and that increasing levels of military expenditures would have decreasing effectiveness as an economic stimulant. The authors predicted that

> it might be wholly impossible to reach a level of full employment by simple increases in the military budget: a bottleneck of specialized scientific and engineering skills could prove to be an insuperable obstacle to further expansion long before the indirect effects of the increased spending had reached the unemployed steel workers of Pittsburgh, the coal miners of Kentucky and West Virginia, the school drop-outs in the slums and ghettos of the big cities all over the country.[10]

Like good Marxists, they were confident of the inevitable crisis and collapse ahead.

Baran and Sweezy were not the first to recognize the pivotal economic position of the military. Radical critics like C. Wright Mills and Herbert Marcuse had accounted for it in their ambitious portrayals of American society, and even those of wholly different political persuasion could argue for it in the high councils of government.[11] But *Monopoly Capital* placed the military-economy link squarely in the midst of a general theory, one having genuine intellectual force. Baran and Sweezy's analysis was discarded by many as just so much Red hogwash, but it was persuasive in intellectual circles and became a touchstone of the military-economy debate.

About the same time, another sweeping economic work was making waves: John Kenneth Galbraith's *The New Industrial State.* First appearing in 1967, the Harvard professor's most important contribution to the dismal science was attacked from the Left and the Right as either a celebration of U.S. corporate and technocratic power, or as a socialist's distortion of American industry and its relation to American government. Galbraith had described, in his typically eloquent manner, how a planning system—consisting of a web of interlocking interests, public and private—had supplanted the market as the key economic factor in American life. And an integral part of the planning system was the military.

Galbraith described a technostructure—the complex of those "who bring specialized knowledge, talent or experience to group decision-making,"[12] mainly in the large corporations—which aims to manage consumer demand. In this pursuit, the technostructure must join forces with government: the latter provides the high educational requirements, research and development labs, the security, the regulation of aggregate demand, protection from the confiscatory sentiments of the populace, and so on. The goals of these major public and private actors merge into security, stability, prosperity, and, of course, maintenance of power. The state is thus an essential part of the planning system. And the military, as the dominant locus of power in the state, is visibly entrenched in the system's foreground. "The Department of Defense," Galbraith wrote, "supports the most highly developed planning in the planning system. It provides contracts of long duration, calling for large investment of capital in areas of advanced technology. . . . For no other products can the technostructure plan with such certainty and assurance."[13]

An important byproduct of this seamless association, Galbraith asserted, was a profound identification of the technostructure and its constituents with the military; it is, moreover, more than simply the "military-industrial complex" at work. The corporation is not only involved in, say, the design of an aircraft, it is also intimately involved in formulating its purpose and, ultimately, the context in which the aircraft is being built and deployed. "If the firm has been accorded a more explicit planning function, it helps establish assumptions as to the strength and intentions of the probable enemy, in practice the U.S.S.R., the nature of the probable attack and of the resulting hostilities and the other factors on which defense procurement depends," Galbraith explained. Thus the defense contractor strongly shapes "the official view of defense requirements and therewith of some part of the foreign policy. These will be a broad reflection of the firm's own goals; it would be eccentric to expect otherwise."[14] This identity is an important source of motivation within the new industrial state, which is not driven by profit alone; the influence accorded the technostructure is its own reward.

Galbraith's depiction, like Baran and Sweezy's, had antecedents, but he was able to describe the military-economy relationship in a new way. Instead of money-grubbing aviation companies or weapons-makers feasting at the Pentagon trough, the technostructure appeared as an ubiquitous, neatly engineered matrix. It encompassed, supported, traded with, and fed from its alter egos in Washington. Consumer demand, national security, and technical development were fine-tuned and modulated by the planning system. No messy market chaos, no uncertainty, no ruinous competition. Defense procurement was the sine qua non of the new industrial state: the biggest, the most amenable to centralized planning, the least accessible to market forces, the most comforting to scientific genius, the most resistant to criticism and scrutiny. It was this persuasive and elegantly simple portrayal that Galbraith contributed. He emphasized not the economy's lust for the military's booty, but rather the centrality of the military's own planning needs, and the military's pivotal place in America's planning system.

Criticisms of Galbraith's view did tend to confine him in a nether world between the then ascendent New Left and market-oriented academic economists. The former decried the absence of a class analysis; the latter resisted his depiction of the corporate half of the economy as a planning system impervious to individual consumer demand.[15] But Galbraith did manage to codify the idea that large, complex organizations—public and private—were the defining characteristic of the U.S. economy, and that planning was the dynamic of these organizations. This view was important in establishing—or, at least, introducing—the notion that the Department of Defense was an immense economic planning agency with profound effects on the economy as a whole and certain aspects of that economy (regional development, technology) in particular. Because Galbraith viewed the planning process with some approval, his opinion about defense economic effects per se was somewhat ambiguous, though he surely preferred planning in support of human services.

Concurrent events began to give the debate a new tone, however: the rapid mobilization for the Indochina war was beginning to concern economists. President Johnson refused to raise a war levy, and the effects of defense spending in an already prosperous economy were viewed warily. At the very least, the "macro" considerations were again prominent; moreover, the assumption of the Pentagon's beneficent impact on growth and employment came under attack along with every other aspect of military operations. Although the inflation caused by the war (which began to appear clearly in 1969) could be explained away by LBJ's unwillingness to succumb to a Vietnam tax, the notion that America "needed" high rates of military spending to bolster its economy was at least suspect. (The Vietnam expenditures, of course, could merely be viewed as an extraordinary event in this respect, a surge that neither proves nor disproves the Baran and Sweezy idea.) There was, however, a school of thought emerging on the Right that did see a direct correlation between war and prosperity: Murray Weidenbaum, then a professor at

Washington University in St. Louis, noted that the stock market rose during wartime, a supporting if inconclusive bit of evidence for the war-prosperity thesis.[16]

Near the war's end, a fresh perspective was offered that would profoundly change the defense-economy debate. That perspective, which aimed to view the specific ways defense procurement interacted with the American industrial system, was forcefully presented by an industrial economist at Columbia University, Seymour Melman.

A longtime critic of the defense establishment, Melman had been working out the particulars of his analysis in several books over the previous decade, but his most compelling argument is found in *The Permanent War Economy: American Capitalism in Decline.*[17] The "state capitalism" that was the result of a growing defense industry had specific attributes and contours that distinguished it from the rest of the civilian economy, and this state capitalism was the key to understanding the operations and impacts of defense procurement. Melman respectfully disagreed with Baran and Sweezy's analysis. Government was not simply the "executive committee of the capitalist class," in the common Marxist phrase; rather, Melman argued, "an economic need for war economy is specific to the state-capitalist parts of the American economy but . . . is not an intrinsic part of civilian economy."[18] Not only does a system of state capitalism differ from the traditional industrial economy, but the influence of the first on the second is not at all monolithic: "War economy does not have a homogeneous effect across the economy but is differentiated in its effects by industry, region, and occupational class."[19]

Melman not only disagreed with the Marxian interpretation in its analytical particulars, he reversed its conclusion. Military spending was not "good" for the U.S. economy in its demand-stimulating role; to the contrary, the DoD procurement process damaged the civilian economy in several different ways. The main consequence permeating the economy is the sacrifice of efficiency. As Galbraith noted, the defense contractor is shielded from all the vicissitudes that should attend a corporation in a market economy: contracts cover long periods of time; costs are recovered automatically, and a profit is guaranteed; cost overruns are expected, indeed, built into the system by the accepted practice of underbidding to secure a contract, and so on. The result, when multiplied by the thousands of contracts managed by the Department of Defense, the Department of Energy (which is responsible for the manufacture of atomic weapons), and the National Aeronautics and Space Administration (NASA, which has several military functions), is a paralysis of inefficiency that cannot but grip the whole industrial empire. As Melman explains:

> The sustained normal operation of a large cost- and subsidy-maximizing economic system produces a major unintended effect in the transfer of inefficiency into the civilian economy. Insofar as the cost-maximizing style of operation is carried with them by managers, engineers or workers as they move individually from military to civilian employment, the civilian economy becomes infected with the standards and practices that these men and women learned in the military sphere. For civilian industry, the introduction of such practices is definitely counterproductive. To be sure, this need not apply to all individuals in the same degree. But to the extent that professional-occupational patterns are transferred, the transfer of inefficiency is "impersonal"—i.e., it operates independently of particular features of individual personality.[20]

Inefficiency is not transferred only by the technicians moving from one sphere to another. It is created by the higher costs endemic to defense contracting, bidding up of raw materials, services, and salaries; the very weight of bureaucratic rules and paperwork that encumber the private companies engaged in military work (or engaged with other firms engaged in military work); and so on. One should not understate the breadth and depth of the Pentagon's influence, Melman warned, for its dominance over the civilian economy was apparent in its "(1) control over capital; (2) control over research and development; and (3) control over means of production of new technical personnel." As a planning system, the military saw to it that it had sufficient funds ($1.5 trillion in "capital" in 1950–75), technology innovation (one-third of all U.S. scientists; 80 percent of federal R&D monies), and new blood (sponsorship of relevant curricula in colleges).[21]

Melman's critique of state capitalism and his exposition of how that system hurts nonmilitary economic activity was a profound turnabout in the debate. Fragments of Melman's analysis had appeared before; the effects on employment, for example, had been calculated more than a decade before. Melman himself had explored many of these same themes in earlier works. But *The Permanent War Economy* seemed to bring it all together: at once it appeared that the old Baran and Sweezy idea, whether in its Left or Right versions, was thoroughly debunked. Melman argued on the grounds of theory and of empirical evidence. His analysis was statistical, anecdotal, historical, and predictive. Most of all, it was stinging. "Traditional economic competence of every sort is being eroded by the state capitalist directorate that elevates inefficiency into a national purpose, that disables the market system, that destroys the value of the currency, and that diminishes the decision power of all institutions other than its own," he wrote in the book's preface. "Industrial productivity, the foundation of every nation's economic growth, is eroded by the relentlessly predatory effects of the military economy."[22]

There were, of course, criticisms of Melman's sometimes passionate tirades against the Pentagon. But by the early 1970s the comforting view that the military's expenditures were an unambiguous blessing for American pocketbooks was no longer universally held. In fact, such a stalwart of the American economic establishment as Arthur Burns—a top economic advisor to Eisenhower and chairman of the Federal Reserve Board in the 1970s—could voice doubts: "The real cost of the defense sector consists," he wrote a few years before *The Permanent War Economy* appeared, "not only of the civilian goods and services that are currently foregone as its account; it includes also an element of growth that could have been achieved through larger capital investment in human or business capital."[23]

TECHNOLOGY ENTERS THE DEBATE

By the time the dust stirred up by Melman had settled, a new consideration was beginning to exert dominance over the debate: the Defense Department's impact on technology Innovation, technological advance, science-based industry, and the like had long been recognized as a locomotive of American success. And, at least since the 1950s, with the military's increasing emphasis on nuclear missilry, the strong interest of war planners in the possibilities of science was often discussed. The link between the military and technological advance seemed natural enough; throughout history, armies had always employed the latest gadget. In post-war America, it was hardly surprising that DoD would seek out technologists, would fund R&D, and would integrate innovations into weapons, communications, planning, and all other operations where new hardware could help. And, indeed, the Pentagon came to be viewed not only as a *user* of technology, but a *creator*: the most striking, but by no means the sole, instance of that being the innovation that ushered in the post-war era—atomic weapons.

By the mid-1970s, though, two factors began to alter the perception of technological development and its relation both to economic growth and the military. The first event was Vietnam. The long campaign in Indochina had employed a dazzling array of technologies in support of the American cause. Although the entire arsenal was not exhausted—nuclear weapons were never used, for example, and sea power was limited by the nature of the war—the sheer quantities and varieties of hardware used (and lost) in Southeast Asia brought home the military's increasing dependence on technology. As the nation's first "television war," Vietnam also displayed this employment of gadgetry to its large living room audience. Moreover, since the U.S. expedition in Indochina was a failure, the pervasive use of technology was all the more intriguing.

The second factor was the sharp decline of the economy itself. Between 1974 and the early 1980s, the nation's economic performance reached several post-Depression records for unemployment, business failures, inflation, high interest rates, idle capacity, and just about every other category of significance. Many explanations have been summoned to account for this decline. The aforementioned inflation spurred by the prosecution of the Vietnam War is a major contributor to the traumas of the last decade. The other prominent reason was the OPEC oil embargo and the subsequent oil crisis of 1973–74, and again in 1978–79, in which oil prices quintupled and then doubled again. Not only did this shock the Western economies and precipitate the severe 1974–75 recession, but it changed the possibilities for technology substitutes for labor. The surges of economic growth in the 1950s and 1960s were due in large measure to the steady increases in productivity, in lowering the labor cost per unit of output, and this productivity was achieved in part by substituting cheap energy for labor. Following the energy price hikes, when this technology-for-labor replacement was no longer so cost-effective, productivity growth ebbed, and with it the U.S. economy.

The guns-*and*-butter approach of the Eisenhower, Kennedy, and Johnson years was not so easy in this new environment of austerity. At the same time, the costs of the Pentagon's fascination with technology were escalating, not only in dollars but in the public's acceptance. A shift in the defense-economy debate then was apparent. It was widely acknowledged that the military's large budgets had some—probably harmful—effect on the economy as a whole;[24] but a new attention was being paid to the effects of technology on the military and the effects of the military on technology.

The notion that too much technology was bad for military preparedness was a startling, and controversial, idea. The thesis was forcefully articulated around the same time by James Fallows in his influential book *National Defense*, and Mary Kaldor of the University of Sussex (England) in *The Baroque Arsenal.* Kaldor pointed to the fundamental "technological conservatism" of the U.S. military, to the "routinisation of innovation":

> "Baroque" technical change consists largely of improvements in a given set of "performance characteristics." Submarines are faster, quieter, bigger, and have longer ranges. Aircraft have greater speed, more powerful thrust, and bigger payloads. All weapons systems have more destructive weapons, particularly missiles, and greatly improved capabilities for communication, navigation, detection, identification, and weapon guidance. . . . While the basic technology of the delivery system has not changed much, such marginal improvements have often entailed the use of very advanced technology; e.g., radical electronics innovations such as microprocessors or nuclear power for submarines, and this has greatly increased the complexity of the weapons systems as a whole.[25]

This emphasis on technology—a very narrow application of technology—changes the defense industrial base toward concentration and overspecialization. "The consequence of this elaborate combination of conservatism and technical dynamism," Kaldor continues, "is what economists call 'diminishing returns': more and more effort is expended for smaller improvements in military effectiveness."[26] The modernization of weaponry leads to aircraft that are too complex for pilots to use, or too overburdened with technological capability to be maintained. As Fallows points out, the high-tech aircraft, tanks, and other sophisticated tools of the military trade require extra time to come out of development and testing into production and use, and, once there, require much greater maintenance (often "in the shop" rather than being ready for combat). Even the vaunted capability (speed, detection, fighting, etc.) is often less than the older machines with less high tech.[27]

Among the effects of this trend toward technical sophistication is, of course, high cost. "The extreme R&D emphasis on equipment performance has rapidly increased the cost of military equipment," writes Jacques Gansler, "and thus greatly reduced the quantities of equipment procured. Performance has been increased, but at too high a cost. By contrast, in the civilian world R&D is used to reduce cost and improve performance simultaneously."[28] Yet this emphasis, Gansler asserts, constitutes the main mission of the defense industrial base—maintaining and enhancing technological superiority. At such prices, fewer and fewer weapons can be developed, and fewer and fewer companies are engaged to produce weapons and accompanying military support systems. As a result, technical innovative processes are more and more brought to bear on a narrow range. Innovation is applied "vertically" (intense improvement of a few "products") rather than "horizontally" (inventing new products and processes).

Overspecialization of the skilled personnel is not the only outcome of this trend. Kaldor describes a "technological overdevelopment," in which a single sector of the economy—be it automobiles or weapons—is granted exceptional resources, and, regardless of how well or far the sector has matured, resources of capital, expertise, production capacity, and so forth, are incessantly made available as a priority to that sector.[29] Because resources are limited, one consequence of this overdevelopment must be to starve other sectors of their due. It also has another effect, one of particular importance to the defense-economy debate: technical overdevelopment, this intense vertical application of resources, stifles innovation within the sector and inhibits commercial applications outside the sector.

Advocates of the military's heavy emphasis on technology typically defend it on two counts. First is the appeal to defense needs. "To respond to Neo-Luddite attacks on high technology," says one Pentagon general, "we must remember that it is the American way of war to extend our technological reach as far as we can, for it is not our national habit to be second-best in battle."[30] Second is the claim that civilian, commercial technology is a natural "spinoff" from military R&D. It is to this latter claim that the notion of "overdevelopment" is particularly relevant.

There is no question that the Department of Defense has played a major role in creating new products, even new industries, as a consequence of military research and development projects. The list often cited is long, and includes aerospace, semiconductors, nuclear power, medical technology, robotics, and many others. The argument is simple: military R&D develops new "products" for defense, but the technology created can also be transformed into something useful for commercial industry. Since the Pentagon is willing and able to underwrite high-risk "blue sky" research and has ample funds to give to promising ideas, it is on the cutting edge of technological advance, actually more risk-taking and adventurous than most private industry. No single institution can so steadily push back the frontiers of knowledge and discovery. And this advance inevitably finds its way into civilian commercial applications. Thus, DoD's emphasis on technology not only enhances U.S. military preparedness, it directly benefits the civilian economy.

The "spinoff" argument is key to understanding the defense-economy debate. If, as Melman and others maintain, the defense establishment is a drain on the civilian sector without returning a sufficient number of tangible benefits, then the implication is clear. But if the return on the defense investment is not only security, but steady advances on the technological front, then the Melman argument is far less compelling.

Actually sorting out the spinoffs in any particular sector is not an easy task, however, and this complicates any assessment of the spinoff argument. Consider, for example, space technology. Following the Sputnik shock of 1957, programs for exploring outer space were initiated simultaneously; the Department of Defense would oversee military uses, and the newly formed NASA would handle civilian space activities. The funding for military space was clearly the prime mover in the federal government's activism, and, naturally enough, the private firms working in one area also tended to work in the other. NASA is credited with building the technical resource base of the national space effort, particularly "in developing the capability of American educational institutions to train technicians and engineers for the rapidly evolving technologies of space."[31] Yet, even though it was the express policy of the space programs, the transfer of technology—especially from DoD to NASA—is considered to have been desultory. Secrecy, genuine concerns about national security, and a lack of incentive to transfer technology from defense to nondefense, have all contributed to the slow pace of technology sharing in the twenty-five years of U.S. space activities.[32] "As the military program has grown in the past decade," a 1983 government report states, the technology-transfer "difficulties have become more common."[33] It is apparent, moreover, that the hope of a profitable commercial space industry (beyond telecommunications to materials processing and other large-scale ventures) now hinges in part on foreign competition. Although the European Space Agency spends only 0.04 percent of its members' combined GNP on space compared with the 0.2 percent the United States commits to space, "the areas in which European technology is commercially competitive with the United States are significant and growing."[34] It is no surprise to learn that Japan is also finding space technology "a major area of interest, and government and industry have been working closely together to prepare for Japanese entry into world markets."[35]

It is apparent that the military's involvement with space has been positive in one regard (enabling funding, forging national consensus on space's importance) and negative in another (screening out valuable information from civilian applications). It is impossible to say with much confidence that the space technology industry would have developed more or less satisfactorily in the absence of the strong Pentagon role. The example of space is not entirely suitable, however, because both military and civilian development were administered by federal agencies. The evolution of the aircraft industry offers a clearer public-private split in its military and civilian manifestations, but the evidence of spinoffs and the resultant health of the industry are no less complex. To some analysts, there would be no aviation industry were it not for the military's deep involvement following World War II. The typical rendition of this line of reasoning is that several aircraft designs spurred by the military later were transformed into commercial airplanes: The Boeing B-47 bomber metamorphosed into the Boeing 707; or that, according to Representative Les Aspin, the Boeing 747 was "based on the losing design for the Air Force C-5 cargo plane."[36] The most obvious retort to this claim is that there exists no compelling reason to believe that successful designs like the 747 would not have been developed without military contracts. Mary Kaldor forwards an additional set of objections. "The war and subsequent dependence on the military market can be said to have stunted the development of the aircraft industry," by shifting it from an innovative mode of design and development to one that stressed "more aircraft per aircraft." The chronically poor profitability of the commercial airline industry is in part due, as one Boeing designer explained, to the fact that the "more an airliner resembles a bomber the less successful it will be."[37] In the meantime, Kaldor contends, "the military market froze the structure of the aircraft industry . . . so that it was very difficult for any individual company to obtain an adequate volume of orders."[38] The process of "technological overdevelopment" then set in—hordes of engineers would design, redesign, and continually overhaul the performance characteristics of each aircraft to meet the Defense Department's exacting criteria. The result, it seems, has been aircraft and airline industries in a perpetual state of crisis, with major firms falling into or near to bankruptcy, and foreign competition now a serious threat.

The "overdevelopment" thesis is essential to the assessment of spinoffs not only in viewing particular industries but in understanding the nature of spinoff potential in general. That is, in the early stage of a technology's evolution, an infusion of cash and other resources from the Pentagon can be an enormous help and indeed can create a broad array of spinoff applications. As a technology goes through the development life-cycle, however, there is tendency for the number of spinoffs to become less and less possible. From the military standpoint, the technology's refinement is geared to making it combat-ready (in military parlance, "ruggedizing" the product to withstand speed stress, extremes of temperature, radioactivity, and so on) or to fulfill other special specifications that stir little commercial interest. Another consequence of overdevelopment is an industrial structure in which a few prime contractors dominate the military market. As economists have long pointed out, one of the worst effects of economic concentration in any industry is that it tends to inhibit innovation and efficiency. In noting that eight corporations received 45 percent of DoD's R&D funds in 1977, Gansler asserts: "These large firms emphasize risk minimization, and thus tend not to push new ideas or applications."[39] (He also notes that military R&D done in national government labs, which is increasing rapidly and accounts for nearly half of defense expenditures in this area, is not done with a view to commercial applications—since government employees may be less sensitive to such possibilities—thus further inhibiting the spinoff effect.) Finally, the procurement process itself diminishes the possibilities for spinoffs: "New ideas from outside the existing defense industrial community are greatly discouraged by government procedures."[40] The paperwork and repetitiveness of staff review alone may convince small firms that the contracts are not worth the remuneration.

Whatever the merits of the spinoff argument, the focus of the defense-economy debate shifted again with the ascension of the Reagan administration in January 1981. Concerns revolved around the

Pentagon's substantial demands on resources, because the new administration's budget escalations were concentrated in procurement and R&D. Indeed, these concerns were expressed quickly and widely. As noted earlier, several "mainstream" economists argued that the rapid defense buildup would surely result in production bottlenecks. Charles Schultze, chairman of the Council of Economic Advisers under President Carter, testified to Congress in 1981 that the "procurement-oriented nature and very rapid pace of the planned defense buildup will place substantial strains upon the relatively limited sector of industry that produces for the defense department." The economy as a whole, which then was mired in a deep recession with considerable slack in capacity, would not be unduly affected, Schultze contended. "But an 80 percent rise in the real volume of military procurement and R&D in the short space of four years will give rise to shortages within the defense industries themselves, of skilled labor and specialized components."[41] Others were less sanguine. Lester Thurow, in a celebrated article, "How to Wreck the Economy," laid out what became known as the neoliberal, or "Atari Democrat," position, citing the inflationary effects of the buildup. Thurow also pointed to the inherent threat in the procurement emphasis to America's trading position:

> Would the typical engineer rather work on designing a new missile with a laser guidance system or on designing a new toaster? To ask the question is to answer it. Military research and development are more interesting since they are usually closer to the frontiers of scientific knowledge and are not limited by economic considerations such as whether a product can be sold in the market. The military is willing to pay almost any premium to have a superior product. The civilian economy is not. As a result the most skilled technicians and scientists move into defense.
>
> But suppose you own a civilian computer firm in Boston and many of your best people leave to work in Boston's higher paying and more exciting aerospace firms. How do you compete with Japanese computer firms that will not be losing their most brilliant employees? The Japanese engineer might also like to work on missiles but he does not have the opportunity to do so. The result is that you cannot compete; the Japanese computer industry could well drive the American computer industry out of business.[42]

The answer to this profound challenge typically revolved around a macroeconomic defense, or the spinoff argument (when it was not simply discarded on the basis of national security). As one military economist articulated the first:

> If defense spending is more capital-intensive, this furthers America's retooling and reindustrialization. And if such spending initially goes to higher income, high-saving groups—well, aren't we trying to stimulate savings? If such expenditures bid up scientists' and technicians' wages, won't this encourage entry into these highly productive, crucial fields?[43]

The "slack capacity" rationale was frequently evoked as well: that the nation's idle factories and unemployed workers would hardly be hurt by the DoD budget's "shot-in-the-arm." The danger, retorted the critics, would not appear immediately, but when the economy was again operating at full throttle; the economic effects of the Vietnam buildup were not visible until a few years later.

Several economists once again resorted to the spinoff rationale in order to discredit the Thurow alarm. Any truth in the Thurow argument, said Charles Schultze, is offset by "the spillover of defense-financed technology into the civilian sector. . . . It is probably no accident that Japanese competitiveness and exports are particularly strong in consumer goods while the U.S. is a strong world competitor in fields more closely related to defense, such as aircraft, computers, large scale communications equipment, and the like."[44] Yet, by the mid-1980s, the warning signs were more troubling than ever. The competition from abroad in computers, airlines, and the other traditional strongholds of American global clout was especially worrisome.[45] A congressional report demonstrated that the civilian R&D budgets of Japan, Britain, France, and Germany had risen in the 1964–79 period from 10 percent of U.S. R&D civilian spending to 121 percent of the U.S. commitment.[46] Between 1980 and 1984, federal support for research and development had declined by 15 percent while military R&D increased 110 percent. A large part of the DoD programs, moreover, have little potential for civilian "spillovers."[47] The most dramatic warning sign, however, was the bottom line itself: a 1983 U.S. trade deficit that totalled $69.4 billion, 62 percent higher than the record set in 1982, with the major "villain" being Japan ($21 billion) and the source of imports being manufactured goods (rather than the 1970s' culprit—oil from OPEC). Nor was the prognosis encouraging. The Reagan administration expected a $100 billion trade shortfall in 1984.[48] Clearly, the trade challenge from our military allies, and the Defense Department's intimate relationship with technological advance, could no longer be so cavalierly ignored.

THE PENTAGON LEVIATHAN

A corollary effect of the Defense Department's reach throughout American society also began to gain notice in the 1970s and 1980s. The pervasive influence of the "military-industrial complex" had been studied and remarked upon frequently since President Eisenhower's famous remark in his 1961 Farewell Address. The relationship, which typically characterized defense procurement as hostage to the defense industries' own sophisticated planning and lobbying, is now widely accepted as a fact of American politics.[49] The popular conception of the military-industrial complex probably holds that the defense contractors have too much influence over Congress and the Defense Department, that the "revolving door" between the Pentagon and the defense industry is swinging freely, and that, in sum, the interests of the nation—particularly the cost to the taxpayer—of this cozy arrangement are not well-served.

Although this perception is certainly accurate in most respects, it misses what may be the more significant characteristic of the equation. That is, as an economic agent, the Pentagon wields more influence than the defense industry, and, indeed, must be viewed as the principal economic influence in the United States. During the decline of American economic power in the 1970s, liberal economists argued that the United States needed to adopt a comprehensive strategy of "reindustrialization," perhaps emulating Japan's Ministry of International Trade and Industry (MITI). The old debate about a "free market" versus "centralized planning" was rehashed, and various forms of "industrial policy" emerged from all points on the political spectrum. What was lost on many of the participants in this debate was that the United States already has a kind of MITI; it certainly has planning for a major share of its economy. And that planner is the Pentagon.

Roughly 10 percent of GNP is directly dedicated to military purposes. One-third of all scientists and technicians work on DoD projects. Nearly half of the non-obligated share of the federal budget is spent by the Defense Department, NASA, and the weapons labs. The federal government is also the single largest consumer of goods and services in the nation, and 85 percent of that purchasing power is exercised by three agencies—DoD, NASA, and the Department of Energy (DOE).[50] The sheer size of the military surely indicates an enormous impact on the economy as a whole.

The analysts who have forwarded industrial policy as good medicine for America's economic ills have spoken with great clarity about the haphazard quality of government policy. Government activism in economic affairs is a given—even pro-market "supply side" advocates are quick to use monetary or fiscal policy to achieve goals. What the industrial policy proponents have recommended is that the federal government's many meddlings be carried out according to a plan that improves competitiveness in world markets and pays heed to domestic problems like unemployment and runaway shops.[51] For the most part, however, the theorists of industrial policy have neglected to see the Pentagon as a major actor in the planning scheme.

The Department of Defense does have an "industrial policy" of sorts, but it reflects the overall observation of federal economic activity as being irrational, poorly coordinated, or simply mindless. The self-realization of its planning role appears to be limited.[52] What is most important about the Pentagon's economic planning, however, is that it fails to combine its security mission with appropriate economic goals. This "economic dissonance" may be the grounds of the next phase of the defense-economy debate.

America's preparedness economy, or "mobilized polity," as Daniel Bell calls it, is accelerated forward by the military's mode of planning. That is, says Bell, "single-purpose planning."

> Most engineers, developers, industrialists, and government officials are single-purpose planners. The objective they have in mind is related almost solely to the immediate problem at hand—whether it be a power site, a highway, a canal, a river development—and even when cost-benefit analysis is used (as in the case of the Army Corps of Engineers) there is little awareness of, and almost no attempt to measure, the multiple consequences (i.e., the second-order and third-order effects) of the new system.[53]

For the military, of course, the single purpose is the security mission: the purchase and deployment of weapons and personnel. Another consideration—and that comprising the primary economic goal—is reducing costs. Larger economic and social goals, which would ordinarily be the raison d'être of an industrial policy, become quite peripheral.

Political goals, however, are not. A major part of Defense Department planning is directed to disburse broadly its contracting dollars in order to gain political support in Congress for its programs. For example, one of the most controversial weapons programs of the last decade, the B-1 bomber, is planned to be built by 5,000 contractors in all 48 continental states. "This geographic spread gives all sections of the country an important economic stake in the airplane," explained one report. "Many contend that the B-1 has survived as a major U.S. weapon system because of the strong constituencies formed by the widespread distribution of money and jobs."[54] The exact same reasoning has allowed the MX missile to prevail. (One can imagine the amount of planning necessary to make sure several hundred members of Congress will get a piece of the action; yet the planning is not carried out to achieve cost efficiencies, much less to accomplish such goals as trade competitiveness, inner-city employment, energy efficiency, or numerous other hopes of government activism.) In this respect then, the "iron triangle" of defense contractors, the Pentagon, and Congress, does exert tremendous influence over policy and planning—indeed, is a main object of planning—and operates through devices as varied as Defense Advisory Boards dominated by the constituents of the defense industrial base to grassroots organizing of employees by the major prime contractors.[55] The flow of defense dollars has also profoundly affected regional development, as a number of studies have shown. A major shift of economic activity to the South, Southwest, and Far West has not only uprooted families and has wasted the "standing capital" of idle plants in the older industrial regions, particularly the Midwest.[56]

So the twin focus on the mission of security and on political goals results in a DoD planning process that must often be harmful to the economy—it is planning for political gain rather than efficiency. An excellent example is that of labor. The macroeconomic studies of defense and employment generally conclude that a billion dollars injected into the economy—from whatever source—is likely to produce the same number of jobs, if not directly then as a consequence of the "multiplier effect." One prominent assessment notes that "additional spending on defense and on nondefense purchases of goods and services appears to have roughly equal expansionary effects on employment in the short run."[57] The views in this debate can range far from that fulcrum. Secretary Weinberger, for example, says that "I don't think we should spend money on defense because it is a good jobs-producing program, but the simple fact of the matter is that it is. These are new jobs that we are talking about."[58] The Congressional Research Service contradicts the secretary's contention, pointing to a somewhat lessened multiplier effect for defense; it also underscores the higher wages and capital-intensity of defense, both of which result in fewer jobs per dollar spent.[59]

In an important way, however, these studies and claims miss a larger point, and are typical of how the defense-economy debate is misconstrued. The Department of Defense is engaged, for example, in a serious program of automating the factory, what is called manufacturing technology, or ManTech. Using computer-aided techniques, ManTech is designed to reduce the labor force necessary to produce defense goods, and thereby increase productivity. Although moderately funded ($141 million in fiscal year 1983, less than 0.1 percent of DoD's entire budget),[60] ManTech can have a very large and varied impact on employment.

The immediate effect, of course, is a net loss of jobs (and, whether intended or not, a weakening of labor unions). ManTech will also accelerate a "de-skilling" process that is already considered to be a serious problem in the United States.[61] Here the effect of the Pentagon Leviathan is potentially profound: the ManTech programs are automating factories and reducing the need for skilled labor at a pace that private industry probably could not achieve, yet the corresponding need for retraining workers is not DoD's responsibility. A parallel phenomenon may be seen at the other end of the labor market. Defense Department demand on scientists and engineers is quickening (and is facilitated with higher salaries), but DoD has a limited programmatic interest in expanding the nation's capacity to produce engineers, physicists, and the like. It is worth noting that the lack of planning itself has motivated the desire to reduce "labor problems." As Jacques Gansler, in taking note of the defense labor market's "extreme instability," explains: "Because of training costs and because of the lack of 'learning,' group unity, and supervisory continuity, such significant labor instabilities, which appear to occur throughout the defense industry, result in great economic inefficiencies."[62]

This one brief example ties together several elements of the defense-economy debate. Here we have a prime area of DoD planning, one in which state-of-the-art technological development is integral, and one that affects the nature of economic activity in the United States quite directly and significantly. Yet, the planning is, again, only of the "single-purpose" variety, with second-order effects largely ignored; the impact on productivity and the ultimate benefits for commercial development are, at this point, unclear; and the final outcome for employment and regional development (with attendant "social costs" that must then be borne by some other government entity) is very likely to be wasteful. At the same time, the conventional characterization of "employment" in the defense-economy debate regards only the number of jobs that a billion dollars might or might not produce.

At the very least, the theorists of industrial policy should take a closer look at the Pentagon's industrial policy.

CONCLUSION

The controversy the Pentagon has engendered over its implicit economic impacts has taken many shapes. Generally, it has followed the contours of the larger arguments over economic direction and military policy in general. What has become apparent, however, is that the Pentagon's impact on the economy, and on technological advance in particular, can no longer be ignored. The American economy is in a chronic state of distress: even in its 1984 "boom," interest rates remain in the double-digit stratosphere, unemployment still tops 7 percent, trade deficits are at record heights, federal budgets are tight, and the extent of poverty is worrisome. At the same time, the Pentagon is assured remarkably high expenditures for the foreseeable future.

Fortunately, a new appreciation for the role of technical innovation in our economic well-being is apparent in the 1980s as well. The high-tech revolution is in full bloom. The importance of the tripartite relationship between military procurement, economic health, and technological advance should also be clear today, but the exact nature of that relationship is still vague. As has been demonstrated, the defense-economy debate could not delineate this vital—perhaps decisive—relationship with its deliberate emphasis on macroeconomic categories and political agendas. Locating and exploring the specific places and people in the defense–high-tech partnership is the surest route to discovering the military's impact on advanced technology.

NOTES TO CHAPTER 1

1. Weinberger, Speech before the Miami Chamber of Commerce, September 15, 1982. Quoted from a Defense Department press release.
2. Congressional Budget Office, *Defense Spending and the Economy* (Washington, D.C.: Government Printing Office, 1983), p. xvi.

3. Karen Arenson, "How Large an Economic Share Will Defense Get?" *The New York Times*, February 6, 1983, section IV. There was also a qualification in that view, and that was the broad concern over deficits, as briefly discussed by the CBO above. A more blunt depiction is given by Jacques Gansler, former Deputy Assistant Secretary of Defense, a defense consultant, and author of *The Defense Industry:*

> Based on defense's nominal impact on overall employment, inflation and productivity, it would appear that the nation can provide this defense growth without it being at the expense of its important, non-defense federal programs, or future economic growth. However, there is a necessary penalty to pay; namely, a tax increase to cover the projected deficits. [*Foreign Policy* (Fall, 1983).]

4. Quoted in Arenson, "How Large an Economic Share . . . ?" Even this "official" admission was made grudgingly. In Secretary Weinberger's *Annual Report to Congress*, Fiscal Year 1984 (issued in early 1983), bottlenecks at the prime contractor level is discounted because there is no major overlap between military and civilian production. "The exceptions to this statement," according to Weinberger, "are lower-tier producers of specialized defense goods and producers of non-specialized goods of which defense production takes a substantial share. It is only in these cases that it would be sensible to look for bottlenecks." Pp. 69–70.
5. Weinberger, Speech before the Miami Chamber of Commerce, September 15, 1982. Quoted from a Defense Department press release.
6. Arenson, "How Large an Economic Share . . . ?"
7. Paul A. Baran and Paul M. Sweezy, *Monopoly Capital: An Essay on the American Economic and Social Order* (New York: Monthly Review Press, 1966), p. 176. Another fine analysis appearing then was H.L. Nieburg, *In the Name of Science* (New York: Quadrangle Books).
8. Baran and Sweezy, *Monopoly Capital*, pp. 207, 209.
9. The authors did not consider the Pentagon's demand to be the sole source of the postwar boom; the "automobilization" that had rescued the economy in the 1920s reappeared along with the new phenomena of suburbanization and large amounts of consumer and business indebtedness. Thus, Baran and Sweezy acknowledge that technological innovation can stop the natural tendency of capitalism to decline and collapse, a view now widely accepted in more moderate versions. This relatively contemporary expression of Marxism would hold that the military's relationship to innovation is a key to economic cycles.
10. Baran and Sweezy, *Monopoly Capital*, p. 214.
11. Arthur Burns, chairman of President Eisenhower's Council of Economic Advisers and chairman of the Federal Reserve Board in the 1970s, is quoted by Richard Nixon in *Six Crises* as having urged large new outlays for defense as a countercyclical measure. See Bruce Russett, "Defense Expenditures and National Well-Being," *American Political Science Review* 76, no. 4.

12. J.K. Galbraith, *The New Industrial State*, 3rd edition (Boston: Houghton Mifflin, 1978), p. 74.
13. Ibid., p. 321.
14. Ibid., p. 323.
15. Galbraith addressed his New Left critics directly in the foreword to the third edition of *The New Industrial State* (1978). In pages xv–xvi he writes:

    > The decisive power in modern industrial society is exercised not by capital but by organization, not by the capitalist but by the industrial bureaucrat. This is true in the Western planning systems. It is true also in the socialist societies. It is why the young American radical is no longer attracted by the Soviet model, for that means exchanging one highly organized bureaucratic system for another highly organized bureaucratic system. It is why he can be attracted to Cuba or to China. These are not bureaucratic—or so bureaucratic. But that is because they are not yet industrialized. When they are, they will be. For organization—bureaucracy—is inescapable in a world of advanced industrial technology—large-volume industrial production.

    Galbraith's replies to his market-oriented critics has been a lifelong activity.
16. Murray Weidenbaum, *The Economics of Peace-Time Defense* (New York: Basic Books, 1974).
17. Melman's other books include, most prominently, *Pentagon Capitalism* (1970) and *Our Depleted Society* (1965). Two anthologies he edited were also influential: *Conversion of Industry from a Military to Civilian Economy* (1970) and *The War Economy of the United States* (1971).
18. Melman, *The Permanent War Economy: American Capitalism in Decline* (New York: Touchstone/Simon and Schuster, 1974), p. 261. It is interesting to see the old intellectual dispute between Marx and Max Weber being played out in Melman's analysis as well, that is, the dynamic of class vs. the dominance of bureaucracy. On page 274 Melman states:

    > The theorists of monopoly capitalism make no allowance for the growth of a new center of economic decision-making, the state management, which is endowed with all the production decision-making characteristics that should identify a capitalist ruling class in the tradition of Marx.
    >
    > A managerial-hierarchical institutional network as big as a military economy has a built in tendency for self-perpetuation and self-expansion. The larger the operation, the more specialized the jobs and the longer its duration, the stronger will be the pressures to continue it.
19. Ibid., p. 288.
20. Ibid., p. 69.
21. Ibid., p. 70.
22. Ibid., p. 11.
23. Arthur F. Burns, "The Defense Sector and the American Economy," in Seymour Melman, ed., *The War Economy of the United States* (New York: St. Martin's Press, 1971), p. 115. Of course, many elements of Melman's analysis had appeared before, in his own writings and elsewhere.

The employment effects of the military budget were explored, for example, in W. Leontief and M. Hoffenberg, "The Economic Impact of Disarmament," *Scientific American* (April 1961).

24. An extensive search for evidence by Russett, "Defense Expenditures," to evaluate the notion of the military's harmful effects leads to the following conclusion on page 767 of the article: "Systematic evidence for this argument, though fragmentary, tends to support it. For example, in multivariate analyses of fifteen industrialized economies from 1960 to 1970, [it was found that there are] significant negative relationships between military expenditure and investment and the rate of growth in GNP." Russet finds less evidence of "tradeoffs" in federal spending between defense, on the one hand, and health and education on the other, since the 1940s, until the Reagan administration.
25. Mary Kaldor, *The Baroque Arsenal* (New York: Hill and Wang, 1981), pp. 19–20.
26. Ibid., pp. 22–23.
27. The F-16 fighter is especially instructive as an example. See James Fallows, *National Defense* (New York: Vintage. 1982). It is of primary interest that this viewpoint is not held alone by neoliberals or leftists. The Heritage Foundation, a right-wing "think tank" in Washington that has had considerable influence in the Reagan administration's policymaking circles, has made roughly the same argument: See George W.S. Kuhn, "Department of Defense: Ending Defense Stagnation," in Richard N. Holwill, ed., *Agenda '83* (Washington, D.C.: Heritage Foundation, 1983), pp. 69–70.

    [A] root cause of military stagnation is conceptual. DoD has long committed most of its development resources and analysis to achieving the most complex battle tasks through technology. Tactics have been driven by technology and—to borrow a basketball analogy—hinge on low percentage mid- and full-court shots at the expense of less difficult and more effective lay-ups and twelve-footers. A number of problems have arisen. Low-percentage shots necessitate high-complexity technology. Complexity greatly increases costs to buy and maintain equipment, and drives down the size of fighting forces while it increases the size of the logistics structure. The highly complex technology is also more prone to break down. Meanwhile, no amount of technological complexity changes the long-shot character of complex battle tasks. Consequently, the more complex (so-called higher performance) weapons suffer from markedly lower kill-rates in actual combat than weapons with simpler applications of the same advanced technology.

28. Jacques S. Gansler, *The Defense Industry* (Cambridge, Mass.: MIT Press, 1980), p. 99.
29. Kaldor, *Baroque Arsenal*, pp. 84 ff.
30. General Paul F. Gorman (U.S. Army), "What the High Technology Edge Means," *Defense 83* (Department of Defense publication, June 1983): 26.
31. National Academy of Public Administration, *Encouraging Business Ventures in Space Technologies* (Washington, D.C.: National Academy of Public Administration, 1983), p. 8.

32. Office of Technology Assessment, U.S. Congress, *Civilian Space Policy and Applications* (Washington, D.C.: Government Printing Office, 1983), pp. 160 ff.
33. Ibid., p. 5.
34. Ibid., p. 186.
35. Ibid., p. 203.
36. Gordon Adams, interview with the author, November 16, 1983. Adams states the view, hyperbolically, that there would be no aviation industry without Pentagon support. The spinoff examples are found in Michael R. Gordon, "Will the Pentagon's Ad Hoc 'Industrial Policy' . . . Ultimately Hamper U.S. Industrial Creativity?" *High Technology: Public Policies for the Eighties* (Washington, D.C.: National Journal, 1983), p. 30.
37. Kaldor, *Baroque Arsenal*, pp. 88–89.
38. Ibid., p. 89.
39. Gansler, *Defense Industry*, p. 101.
40. Ibid., p. 106.
41. Charles Schultze, Statement before the Subcommittee on Economic Goals and Intergovernmental Policies, Joint Economic Committee, U.S. Congress, October 13, 1981; manuscript, p. 12.
42. Lester Thurow, "How to Wreck the Economy," *The New York Review of Books*, May 14, 1981.
43. Donald L. Losman, "Defense Helps the Economy," *The New York Times*, November 16, 1982. Losman is a professor of economics at the Industrial College of the Armed Forces of the National Defense University.
44. Schultze, op. cit., p. 5.
45. Schultze's remark, though typical of spinoff aficionadoes, looks to the past rather than the future. The Japanese, to cite the most prominent foreign threat, are reported to be interested not only in aerospace, but in building an aviation industry as well. If not cars, why not planes? As to computers, it is generally acknowledged that the Japanese now build a superior semiconductor; and, while they have mastered "consumer products," they are also pursuing basic R&D and will compete in large-scale communications technologies as well. Consider the view of John A. Young, president of Hewlett-Packard:

    > Japanese government support for technological achievement is unwavering. In a report dated March 1980, Japan's Ministry of International Trade set forth a national goal for the 1980s "to develop innovative and original technology." Pursuing that goal, the Japanese have formulated ambitious plans to capture a major share of the worldwide computer market by 1990 with an advanced fifth generation computer design now under development.

    See John A. Young, "An Agenda for the Electronics Industry," in James Botkin et al., *Global Stakes: The Future of High Technology in America* (Cambridge, Mass.: Ballinger Publishing Co., 1982), p. 174.
46. Cited in Philip J. Hilts, "A Big Military Seems to Mean Less Spending on the Sciences," *The Washington Post*, May 31, 1983.

47. "Defense R&D Grows to Become 67% of all Federal R&D," *Physics Today* (May 1983): 65.
48. "1983 Trade Deficit Hit $69.4 Billion," *The New York Times*, January 28, 1984, p. D1.
49. See, for example, Gordon Adams, *The Politics of Defense Contracting: The Iron Triangle* (New Brunswick, N.J.: Transaction Books, 1982); or Adam Yarmolinsky and Gregory D. Foster, *Paradoxes of Power: The Defense Establishment in the Eighties* (Bloomington: Indiana University Press, 1983). The Yarmolinsky book is an update of his 1971 version, *The Military Establishment.*
50. Ira C. Magaziner and Robert B. Reich, *Minding America's Business: The Decline and Rise of the American Economy* (New York: Vintage, 1982), p. 224.
51. Ibid., p. 255.:

> The United States has an irrational and uncoordinated industrial policy that is comprised of "voluntary" restrictions on imports, occasional bail-outs for major companies near bankruptcy, small sums spent for job training and job relocation, a huge and growing program of defense procurement and defense-related research, and a wide array of subsidies, loan guarantees, and special tax benefits for particular firms or industries. It is an industrial policy by default, in which government and business are inextricably intertwined but in which the goal of international competitiveness has not figured.

Note the difference from Galbraith's depiction of the "planning system" fifteen years earlier.

52. The level of consciousness appears to have reached the plateau of weapons cost reduction, greater efficiency in preparedness, and reducing civilian influence. There is some growing awareness of DoD's planning role and the economy, however. See Philip S. Kronenberg, ed., *Planning for U.S. Security* (Washington, D.C.: National Defense University Press, 1981); and *Defense Procurement and Economic Development* (Washington, D.C.: Defense Office of Economic Adjustment, 1983). Gordon Adams says that at DoD there "are elements of a plan, but not a conscious, overall economic plan. DARPA's VSHIC (very high-speed integrated circuit) program is the closest thing to it, along with DoD's emphasis on manufacturing technology. There is also the Defense Production Act of 1982, which has a job retraining program." Interview with author, November 16, 1983.
53. Daniel Bell, *The Coming of Post-Industrial Society* (New York: Basic Books, 1973), pp. 284–85. Bell's description of the preparedness economy (pp. 355–56) is fascinating:

> It was the root idea of Saint-Simon, August Comte, and Herbert Spencer, the theorists of industrial society, that there was a radical opposition between the industrial spirit and the military spirit. The one emphasized work, production, rationality, the other display, waste, and heroics. . . .
>
> The irony is that although the economizing spirit—the deployment of limited resources to attain maximum results—has indeed spread throughout society as

> Schumpeter and others have argued, war rather than peace has been in large part responsible for the acceptance of planning and technocratic modes of government. . . .
>
> A mobilized society is one in which the major resources of the country are concentrated on a few specific objectives defined by the government. In these sectors, private needs are in effect subordinated to the mobilized goals and the role of private decision is reduced almost to nothing.

Bell attributes rationalization and technocratic planning to have arrived at the Pentagon with Robert McNamara, Secretary of Defense from 1961 to 1967. Nearly all commentators credit McNamara with this achievement—cost accounting related to strategy—but his technocratic thinking apparently did not extend to what we now call industrial policy.

Bell makes one other point worth recalling, especially in relation to Baran and Sweezy: "In one sense, as Herman Kahn has pointed out, military technology has supplanted the 'mode of production,' in Marx's use of the term, as a major determinant of social structure" (p. 356). If so, the impact of the military on America is far greater than this book even implies.

54. "Servants or Masters?" *The Los Angeles Times*, July 10, 1983.
55. Gordon Adams, *The Politics of Defense Contracting*, esp. pp. 7–11, for an excellent summary.
56. Jacqueline Mazza and Dale Wilkinson, *The Unprotected Flank: Regional and Strategic Imablances in Defense Spending Patterns* (Washington, D.C.: The Northeast–Midwest Institute, 1980). One can argue that DoD funds simply follow aerospace and other high tech firms, or, equally plausibly, are part of the political gamesmanship of the Iron Triangle: congressional committee chairmen have traditionally come from the South, for example, and DoD needs to satisfy them with contracts in their districts.
57. Congressional Budget Office, *Defense Spending*, p. 43.
58. Quoted in "Defense Spending: Its Effect on Jobs," *The New York Times*, October 16, 1983.
59. Carolyn Kay Brancato and Linda LeGrande, *The Impact of Defense Spending on Employment* (Washington, D.C.: Congressional Research Service, 1982).
60. Gordon, "Will the Pentagon's Ad Hoc 'Industrial Policy' . . . ?" See also Office of the Undersecretary of Defense for Research and Engineering, *Report of the Defense Science Board 1980 Summer Study Panel on Industrial Responsiveness* (Washington, D.C., 1980).
61. See Pat Choate, "Technology and the Workers," *High Technology: Public Policy for the Eighties*, pp. 88–91. Also see the excellent discussions on education and high tech in Botkin, ed., *Global Stakes.*
62. Gansler, *The Defense Industry*, pp. 50–51.

2 HIGH TECHNOLOGY, DEFENSE, AND INTERNATIONAL TRADE

Robert B. Reich

The United States is engaged in two global contests. The first, a political contest with the Soviet Union, began shortly after World War II; it involves complex weapons systems and rests on a delicate set of alliances and spheres of influence. The second, an economic contest with Japan, began about a dozen years ago; it involves complex technologies and depends on an intricate set of trading relationships. Both contests are critical to the future of America—the second quite as much as the first. But the Reagan administration seems quite incapable of viewing international relations in any terms other than cold war diplomacy.

The fiasco over the Soviet gas pipeline is a recent example. In October 1981, after six years of negotiation, the Western European nations agreed to supply the Soviets with equipment to build a pipeline that would carry natural gas 3,600 miles from the Siberian fields to Western Europe. The equipment was to include compressors, which would work like giant fans to push the gas along the pipeline, and some 125 turbines, which would supply the power to operate the compressors. The most intricate part of a turbine is the rotor—a collection of carefully shaped blades, arranged along the shaft. Their manufacture requires sophisticated casting techniques and exotic metals. The rotors were to be designed and supplied by America's General Electric Company; equipment for the compressors was to come from other American companies such as Dresser, Caterpillar, and Cooper Industries. But on December 11, 1981, citing the Soviet Union's "heavy and direct responsibility for the repression in Poland," the Reagan administration abruptly barred American exports of any high technologies to the Soviets, forcing the cancellation of these subcontracts, which were worth some $250 million.

The December 11 ban did not have much impact on the timetable for the pipeline, however, since several European companies, which hold licenses from the U.S. firms to manufacture American-designed equipment on their own, stepped in to fill the gap. Even this inconvenience appeared to be short-lived when, at the Versailles economic summit in June 1982, then Secretary of State Alexander Haig suggested to Western European leaders that the president would ease the pipeline restrictions if the Europeans would limit their practice of extending favorable credit terms to the Soviets. The Europeans accepted the bargain, agreeing to raise their interest rates on Soviet loans. But then the other shoe dropped: on June 18, 1982, President Reagan, apparently irked by statements by some Europeans that the limits on export credits were meaningless, extended the original pipeline ban to cover sale to the Soviets by foreign subsidiaries of American companies and also to include sales by foreign firms of American-licensed technology. The reason given for this sweeping ban was that the Soviets had simply failed to respond to the earlier sanctions.

European reaction was swift. The French and Italian governments promptly instructed their companies to defy the American ban and fulfill the Soviet contracts. West German Chancellor Helmut Schmidt urged German companies also to flout the U.S. sanctions, and there is no indication that the new chancellor, Helmut Kohl, disagrees. Even Margaret Thatcher, Reagan's ideological soulmate, invoked Britain's 1980 law on trade, which limits the extraterritoriality of United States laws. On August 25, 1982, the French subsidiary of Dallas-based Dresser Industries began loading three large compressors onto a cargo ship bound for the Soviet Union. Other European companies have followed suit.

The pipeline simply means too much to the Europeans to bow to American pressure. It promises jobs for several thousand European workers, orders worth at least $11 billion, and a source of energy from somewhere other than the Persian Gulf. And even if the pipeline ban could be enforced, it would not seriously hurt the Soviets. At most, it would postpone the building of the pipeline for one or two years while European companies redesigned the turbines and the Soviets produced substitute rotors.

The administration contends that the pipeline will make the Europeans overly dependent on the Soviets and therefore render them more vulnerable to Russian power politics. By the most optimistic forecast, Russian gas exports to Western Europe will account for less than one-fourth of Europe's gas consumption by 1990—with much of the rest still coming from the Middle East. And that fraction of Europe's gas needs will amount to a much smaller fraction—no more than 8 percent—of Western Europe's total energy bill. From the standpoint of Europe's vulnerability, it seems wiser to diversify energy supplies than to continue to gamble on a steady supply from the Middle East. More to the point, with its missiles poised at Europe's major cities and its tanks amassed along Europe's borders, the Soviet Union has more effective means for twisting Europe's collective arm than sacrificing revenues on its sale of natural gas. The leaders of Western Europe may not wear Adam Smith ties, but they do understand that because trade generates benefits for both traders, it inevitably creates *mutual* dependencies, and this mutuality deters either side from doing anything that may displease the other. The Europeans see the pipeline as a bulwark *against* Soviet power politics in the region.

Europe's open and noisy defiance of the pipeline ban is a blow to American credibility and a setback for the Atlantic alliance as a whole. But there is no serious danger that this break signals the beginning of a gradual dissolution of NATO. Europe and America simply have too much at stake in their mutual security to let a pipeline come between them. The real damage of this imbroglio will be to the American economy, and the real beneficiaries will be the Japanese. To understand why this is so, it is necessary to look at the new global environment in which Japanese and American companies are now competing.

High technology components are the fastest growing and most competitive segments of international trade. They are coming to be the building blocks for countless manufactured products. Precision gadgetry like semiconductors, microprocessors, lasers, fiber-optic cables, robots, turbines, and rotors are finding their way into all sorts of complex machines, from automobiles to guided missiles. Because semiconductors can store huge amounts of information on the miniature circuits that are etched into them, for example, they are key elements in the new computer, telecommunication, and aerospace technology. Total world sales for these wafer-thin, fingernail-sized components have more than quadrupled since 1970. From $10.5 billion in 1981, they will reach $80 billion by 1990. Dominance in technological building blocks like these will provide the same economic strength that steel production gave the United States in the first half of this century. The nation that can produce them cheaper and better than any other will have a huge advantage in producing and selling the advanced technologies in the future.

Success in selling such components as miniature circuits and fiber-optic cables depends on volume, experience, and technological innovation. Since much of the technology can be reproduced quite simply, the cost of making it declines rapidly with the volume of production—a 15 to 25 percent drop every time the volume doubles. With experience in reducing the cost of material and overhead, and in fabrication, the cost can drop still further, while quality improves. Even the production of precision products like rotors benefits substantially from know-how and experience.

The racecourse is worldwide. Japanese and American companies are competing to obtain volume and experience by selling around the globe to manufacturers that use the technologies in their final products. These Japanese or American companies sometimes gain more ground by licensing a foreign company to manufacture the equipment on its own, on the basis of Japanese or American designs; this is typically done when there are substantial costs to transporting the equipment, or where a foreign country—anxious to preserve employment and gain technological know-how—has erected import barriers. The American or Japanese company that licenses the technology earns royalty fees that help it pay for further research and development, and also gains potential customers for additional technology that complements the licensees' equipment. In this way, licensing often provides a foot in the door to obtaining experience and volume in a whole range of related components.

Viewed in this light, the pipeline ban appears more ominous. The real losers are American companies that supply or license European manufacturers with high technology components. In the future, these European manufacturers will think twice before contracting with an American company. Who knows when another ban—tenuously related to American defense interests—will be invoked by the White House? Whenever possible, these European manufacturers will look to Japan for their high technology needs. They have learned their lesson, and American companies—not just General Electric, Dresser, Caterpillar, and Cooper, but a whole array of American high technology manufacturers—will surely suffer for it.

The same sort of economic myopia can be seen in the Reagan administration's policies toward Latin America. At the October 1981 Cancun conference in Mexico, the administration piously decried the use of foreign aid and export credits to bolster the economies of our southern neighbors, preferring to rely instead, as President Reagan phrased it, on "the magic of the marketplace." It was to be through trade, not aid, that Latin Americans would come to embrace the wonders of capitalism. In retrospect it is clear that this talk was aimed more at shoring up America's political and military influence south of the border than at bolstering the region's economies. While arms sales increased 30 percent in 1982, the door to the promised U.S. market has been slammed shut. In May 1982, Reagan signed a proclamation imposing quotas on sugar imports for the first time since 1974. As a result, Latin American and Caribbean exporters stand to lose some $180 million annually. Other Latin American industries are also being denied access to the American market. Small wonder that Latin America, in turn, has relatively little interest in paying high prices for American commercial high technology.

Meanwhile, the Japanese are busy building markets in Latin America for their high technologies. The Japanese understand that the demand for many products that incorporate high technologies is growing faster there than in industrialized countries. For example, sales of automobiles, television sets, and home appliances are sluggish in the United States and Western Europe, both because of the recession and because most Americans and Europeans already own these products. But in Latin America sales of the same products have mushroomed in recent years, and once the current world recession is over, we can expect further growth. By selling in these expanding markets, building manufacturing facilities there, licensing Latin American producers to manufacture Japanese-designed components, and providing Latin Americans with entire plants of their own, the Japanese are participating directly in that growth. Japanese companies thereby gain the sales volume they need to set a very low price for their high technologies, enabling them to undercut American competitors even in the U.S. market. At the same time, the Japanese are setting up channels to market their older technologies.

In this way, Latin America is being incorporated into the strategies by which Japanese companies plan to grow. Meanwhile, Latin Americans are gaining the resources and skills necessary to make use of the new technologies. All this is made possible by the continual forward movement of Japanese industry. Collaborating with the government, Japanese companies are willing to discard older technologies as fast as newer ones can be developed, while financing the development of the new technologies by gaining strong and sometimes dominant positions in the world market for older ones. Japan's Overseas Economic Cooperation Fund provides very low-interest loans to Latin Americans to finance large technological purchases, particularly of whole manufacturing plants (60 percent of Japan's "Ex-Im" loans are aimed at selling whole plants). Japan's tax laws provide additional incentives for technological transfer, and Japan provides its companies with generous insurance against foreign losses. American foreign policy, obsessed with military and diplomatic advantage, is blind to this dynamic competitive process.

The same failure to understand international competition is affecting the very development of American high technology. The Pentagon now funds about 30 percent of all the research and development undertaken in the United States—almost twice the proportion as funded by the Japanese government. For basic research, concerned with broad-based and theoretical experimentation that may have few immediate commercial applications, government funding exceeds two-thirds of the total.

Right now the Pentagon is funding research and contracting for very large-scale integrated circuits (the next major stage in the evolution of semiconductors), computer-aided manufacturing technologies appropriate to a "factory of the future," advanced fiber optics, lasers, and a "fifth generation" computer. Japan's Ministry of International Trade and Industry (MITI) is pushing the same technologies. But unlike MITI, the Defense Department has no interest in the successful marketing of these new technologies. The Pentagon wants new and ever more advanced weapons systems. The two goals have begun to diverge sharply.

The marketing of new commercial products is stimulated by domestic competition, which forces firms to improve their performance and aggressively seek foreign outlets. Although MITI allows firms to cooperate on specific basic research projects, it ensures that they are fiercely competitive in marketing. For example, thirty-two Japanese companies now produce semiconductors, and the competition is intense. But the Pentagon is unconcerned about competition within American industry. Over 65 percent of the dollar volume of U.S. defense contracts is now awarded without competitive bidding. And even where competitive bidding occurs, the bids are often rendered meaningless by large cost overruns. The Pentagon seems most comfortable with large, stable contractors who are relatively immune to the uncertainties of competition.

Marketing new products successfully also requires long lead times, during which firms can apply new technologies and make sure they have adequate capital, labor, and productive capacity to meet anticipated demand. Many MITI projects span a decade or more. But Pentagon programs are subject to relatively sudden changes in politics and in perceptions of national security needs. The precipitous rise in U.S. defense spending planned for the next five years is likely to create bottlenecks in the production of key subcomponents and capital goods, and shortages of engineers and scientists in advanced electronics and machinery. Even in the midst of a recession, unfulfilled defense orders totalled $63 billion in 1981—up 30 percent from 1980. And there is already a shortage of an estimated 60,000 skilled machinists.

Finally, and most important of all, commercialization requires that new technologies be transferable to commercial uses at relatively low cost. MITI sees to it that new technologies are diffused rapidly into the economy and incorporated into countless commercial products. But the advanced designs required by tomorrow's elaborate military hardware—designs incorporated into precision-guided munitions, air-to-air missiles, cruise missiles, night-vision equipment, and missile-tracking devices—will not be as easily applicable to commercial uses as were the more primitive technologies produced during the defense and aerospace programs of the late 1950s and early 1960s. Indeed, it is precisely because America's commercial high technologies are not likely to be adaptable to defense needs in the years ahead that the Defense Department has launched its own research and development programs to produce advanced gadgets designed expressly to meet its own needs. Rather than encourage American commercial development, defense spending on emerging high technologies will have the opposite effect over the long term, diverting U.S.

scientists and engineers away from commercial applications. And Pentagon jitters about leaks to the Soviets are casting a veil of secrecy over commercial high-tech research (at one recent photo-optics engineering meeting, 170 unclassified papers had to be withdrawn). The big losers: American entrepreneurs scouting for new inventions.

This disjuncture between defense needs and the development of commercial high technology is likely to loom larger in the next few years, as the defense buildup proceeds apace. Planned military spending will exceed $1.5 trillion over the years 1982–1987. This will profoundly affect several emerging industries. Between now and 1987, for example, defense spending for semiconductors is expected to increase by 18.3 percent, while commercial semiconductor purchases will increase by only 11.8 percent. A similarly divergent growth pattern is expected for computer sales (16.4 percent for defense, 11.8 percent for commercial purposes), engineering and scientific equipment sales (9 percent for defense, 5.6 percent for commercial purposes), and sales of communications technologies (11.6 percent for defense, 5.3 percent for commercial purposes). There are already signs that the Defense Department is taking over the space shuttle program, thereby cutting off the shuttle's fledgling commercial applications. At the same time, the department's share of government research and development outlays is expected to rise to over 60 percent by 1983.

The only concerted effort by Washington to meet the Japanese threat in commercial high technology is through veiled attempts at sealing off the American market from the Japanese. In early 1982, for example, the Defense Department and the Federal Communications Commission both urged AT&T not to award a large fiber-optics contract to Fujitsu, even though the Japanese company had offered the lowest bid. Obediently (and with respectful awareness of the fact that AT&T's fate now rests with several government officials engaged in redesigning communications policy), AT&T has now announced that the lucrative contract, and all future fiber-optic contracts, would be awarded only to American firms. And in recent weeks U.S. prosecutors have launched a spate of legal attacks on Japanese manufacturers: employees of Hitachi and Mitsubishi have been indicted for allegedly seeking to steal computer secrets from IBM; Mitsui, the large trading company, has been charged with selling Japanese steel to Americans at unfairly low prices (Mitsui pleaded guilty to the charge); and the Justice Department is investigating six Japanese semiconductor manufacturers for alleged price fixing. Regardless of the underlying truth of these charges (it would not be surprising if Japanese companies, which exist in a far less litigious culture than ours, fail to observe the niceties of American law), there is little doubt that the purpose and effect of these lawsuits is to deter the Japanese from aggressively marketing their products in the United States.

In all these ways—and in countless others—the United States is sacrificing the nation's economic future to the short-term exigencies of national defense and the parochial demands of domestic producers. America continues to regard the rest of the world through the foggy lenses of cold war diplomacy rather than through the clear glasses of commercial competition. Our international economic policies consist almost entirely of trade embargoes, tariffs, quotas, dumping complaints, antitrust challenges, and an occasional sensational arrest for alleged theft of trade secrets. All this, coupled with a draconian monetary policy that keeps interest rates at record levels, forces the dollar up in international trade, making American goods even less attractive in the global market and cutting short economic recovery in every other industrialized nation. Meanwhile, rather than encouraging our emerging industries and nurturing our high technologies, we are distorting their growth through exorbitant defense expenditures on esoteric military hardware.

It is ironic that the Reagan administration, whose laissez faire rhetoric comes to us almost intact from the nineteenth century, should stymie free and robust international trade and cripple American industry in the process. But the conservative mind still sees the world as a vast chessboard on which subtle games of power politics are to be played—another vestige of the nineteenth century. The international economy is of secondary importance. Foreign policy is the bailiwick of the State and Defense Departments; the Commerce Department plays third fiddle. We have no equivalent of Japan's MITI, which is concerned primarily with the future of the national economy and its place in a changing world economy.

This international myopia is having grave consequences. America is in danger of losing the high technology race. The nation's economy is not evolving rapidly enough for American companies to capture a significant share of the world's emerging markets. Already, the Japanese have 60 percent of the world market in 64k memory chips. They are gaining significant market shares in fiber optics, communications equipment, sensing devices, and composite materials. They are substantially ahead in robotics, computer-aided manufacturing, and photovoltaics. America's declining competitiveness in these emerging technologies will be accompanied by a decline in our standard of living. Indeed, the decline has already begun.

Some will say that all this is well and good. Americans have lived too high on the hog for too long, and it is fitting and right that other industrialized nations should reach and surpass us. Who cares about high technology anyway? The problem is twofold. First and most obviously, if our national economy is no longer growing and many Americans therefore come to feel poor relative to what citizens in some other countries now have, it will be harder than ever to convince them to share their wealth with their less fortunate fellow citizens. All too often, history teaches us, a society's capacity for compassion and civic virtue exists in direct proportion to the rise in its citizens' real incomes.

The second problem with a declining position in international competition brings us full circle, back to national defense. For our ability to maintain peace and deter aggression depends on our overall prosperity; the resources and commitments that national defense requires over the long term can be sustained only amid a growing and bouyant economy. Perhaps even more to the point, America's best guarantee of national security over the long term is a bouyant world economy in which the fruits of prosperity are widely shared. Trade embargoes, tariffs, quotas, dumping complaints, arms sales, and all the other ways in which American foreign policy distorts international trade add nothing to the real wealth of the world. Unlike the military and diplomatic contest that preoccupies the Reagan administration, the context in commercial high technology, at least in some respects, pays a dividend to the rest of the world in the form of a higher quality of life.

The United States can meet the Japanese challenge in high technology only through policies calculated to spur American high-tech producers to commercial success in world markets: generous funding of commercial research and development; insurance against foreign losses; low-interest loans to less developed nations to finance their technological purchases; education loans and grants to ensure an adequate supply of engineers and teachers; awards of defense contracts to smaller, innovative high-tech companies; more defense contracts channeled toward generic technologies with commercial applications; and an open world-trading system that eschews embargoes and import barriers. In short, we need an affirmative industrial policy for American high technology. By taking precisely the opposite tack in each of these areas, the Reagan administration is threatening our economic future, and, in the process, jeopardizing our national security.

3 MICROELECTRONICS AND WAR

Frank Barnaby

In the past fifteen years or so, the characteristics of major weapons—tanks, combat aircraft, missiles, and warships—have changed beyond all recognition. This change is largely due to developments in microelectronics. Military microelectronic systems have revolutionized the guidance and control of weapons, and military communications, command, and intelligence.

It would take many volumes just to describe all the military uses of microelectronics. Consequently, we will discuss here only four applications—in strategic missile accuracy in strategic antisubmarine warfare, in airborne warning and control system aircraft, in the automated battlefield, and in modern combat aircraft. These applications are especially chosen to indicate the range and scope of the military applications of microelectronics and to show the wide ramifications such applications can have for military tactics and strategy, for the future battlefield, and for the development of nuclear policies.

Military technological revolutions that have been, or will be, brought on by microelectronics are most usefully seen in the context of advances in military technology in general. These advances are made possible by military research and development which impels military technology and is, therefore, the activity most responsible for the arms race.

MILITARY RESEARCH AND DEVELOPMENT

If there were no military research and development, no new major weapons would be produced and no significant improvements would be made in the performance of existing weapons. The arms race, at least in the qualitative sense, would soon grind to a halt, even though the size of arsenals may increase and the arsenals of the smaller powers may be brought technologically closer to those of the great powers by global arms trade.

A huge amount of money is spent on the research and development of military weapons: today, a large fraction of it goes into the development of electronics for weapons systems. During the 1960s, an average of about $16 billion a year was spent on military R&D, about 10 percent of world military expenditure. Now, roughly $50 billion a year is spent on military R&D, also about 10 percent of world military expenditure. Even these huge sums, however, are conservative estimates.

In constant prices, in order to take inflation into account, world military research and development spending increased by about 60 percent during the 1960s and by about 20 percent during the 1970s. Judging by the increases in military spending planned by some major countries, military R&D spending is likely to increase more rapidly during the 1980s than during the 1970s.

The money spent by governments on military R&D is a large fraction of the money they spend on research and development in general. In the United States, and almost certainly in the Soviet Union as well, over one-half of total government-financed research and development is invested in the military. Worldwide, about 40 percent of research expenditure is devoted to military research.

About 400,000 of the world's most highly qualified physical scientists and engineers work on military R&D; this number is about 40 percent of the world's research scientists and engineers. If only physicists and engineering scientists are included, the percentage is even greater—well over 50 percent.

For the past two decades the bulk of military research and development has been performed by the United States and the Soviet Union. These two countries account for about 85 percent of the money spent on this activity. France, the United Kingdom, the Federal Republic of Germany, and China together spend about 20 percent as much as the United States. The rest of the world accounts for no more than 5 percent of the total spent on military R&D. Of this group, the most significant spenders are Australia, Canada, India, Italy, Japan, the Netherlands, and Sweden. The amount spent by countries belonging to the Warsaw Treaty Organisation on military R&D is not publicly known. The countries spending the most on military R&D are, in general, also making the most rapid advances in military electronics.

NUCLEAR WAR FIGHTING WEAPONS

The most dramatic consequences of advances in military electronics are those that make a nuclear world war more probable. Such advances include, in particular, improvements in the accuracy and reliability of strategic missiles, but also include the development of space-based navigational aids, improvements in antisubmarine warfare techniques, and the development of methods to destroy enemy ballistic warheads and enemy satellites in space.

Accurate and reliable ballistic missiles are seen as suitable weapons for the fighting of a nuclear war. For many years now, a large fraction—probably more than half—of strategic nuclear warheads has been aimed at military targets, even though these may have been large-area targets, often in or near cities. But with the advent of more accurate warheads, smaller (and, therefore, a greater number of military sites can be targeted. Even though many strategic nuclear weapons have been aimed at military targets, official nuclear policies were based on mutually assured destruction in which the enemy's cities became hostages. The theory was that the enemy would not attack if cities and industry could be destroyed in retaliation. Moves toward a nuclear war fighting strategy are being made not because the requirements of nuclear deterrence have changed (the psychology of the enemy is, after all, the same), but because military technology has made nuclear war fighting weapons available. Once available, these weapons are usually deployed. Policies then have to be modified so that politicians can justify this deployment.

The more the two great powers adapt to nuclear war fighting doctrines, the greater the probability of a nuclear war, since the perception that such a war is "fightable and winnable" will rapidly gain ground. Offensive and defensive strategic nuclear weapons systems may in due course be developed that make a preemptive surprise attack possible, or, in the opinion of some scientists, probable or even inevitable. The likelihood of a nuclear world war due to accident or miscalculation, either because of technological error or because of a misinterpretation of data, is also considerably increased. The more the two superpowers rely on computers to warn them of nuclear attack and to launch and control their nuclear weapons, the greater the danger of nuclear war. Nuclear war is thus becoming increasingly hair-trigger.

Improving Strategic Missile Accuracy

The accuracy of, for example, the current U.S. Minuteman III, the world's most sophisticated intercontinental ballistic missile (ICBM), is being upgraded by improvements in the missile's computerized guidance system. These improvements involve better mathematical descriptions of the in-flight performance of the inertia platform and accelerometers, and better prelaunch calibration of the gyroscopes and accelerometers. With these guidance improvements, the circular error probable (CEP), or the radius of the circle within which a weapon is capable of striking, of the Minuteman III will decrease from the current value of about 350 meters to about 200 meters. Minuteman III ICBMs with the higher accuracy will be able to destroy Soviet ICBMs in silos hardened to about 4,000 pounds per square inch, with a probability of about 78 percent for one shot and about 95 percent for two shots.

The Americans are also developing an exceedingly accurate new ICBM—the MX—that may be the ultimate in ballistic missile design. The guidance for the MX missile will probably be based on the advanced inertia reference sphere, an "all-attitude" system that can correct for movements of the missile along the ground before it is launched. A CEP of about 100 meters should be achieved with this system. If the MX warhead is provided with terminal guidance, using a laser or radar system to guide the warhead to its target, CEPs of a few tens of meters may be possible.

An American strategic ICBM force of this accuracy would be seen by the Soviet Union as considerably threatening its own ICBM force. The most likely Soviet response to this threat would be the installation of a "launch-on-warning" system in which a computer would be used to launch Soviet ICBMs while the American missiles were in flight. A satellite system would detect the American ICBMs as they crossed the horizon, and a satellite signal would then trigger off the computerized launching procedures for the Soviet ICBMs. Initiation of a nuclear holocaust by computer, without any human decision, must surely be the ultimate madness. Even so, it is within sight.

Improvement of Strategic Submarine-launched Ballistic Missiles. The Soviet and American navies operate a total of 113 modern strategic nuclear submarines—the Soviet Union with 72 and the United States with 41. The ballistic missiles carried by these submarines are normally targeted on the adversary's cities to provide the assured destruction on which nuclear deterrence depends. A single modern U.S. strategic nuclear submarine, for example, carries about 200 nuclear warheads, enough to destroy every Soviet city with a population of more than 150,000. American cities are hostage to Soviet strategic nuclear submarines to a similar extent. Four strategic nuclear submarines (out of the 113) on appropriate stations could thus destroy most of the major cities in the Northern Hemisphere.

The quality of strategic nuclear submarines and the ballistic missiles they carry is being continuously improved. In the United States, for example, the present Polaris and Poseidon strategic nuclear submarine force is being augmented, and may eventually be replaced, by Trident submarines.

Trident submarines will be equipped with a new submarine-launched ballistic missile (SLBM), the Trident I, the successor of the Poseidon C–3 SLBM. Yet another SLBM, the Trident II, is currently being developed for eventual deployment on Trident submarines. In the meantime, Trident I missiles will also be deployed on Poseidon submarines.

The circular error probable of the Trident I SLBM is about 500 meters at maximum range, whereas that of the Poseidon SLBM is about 550 meters. The development and deployment of mid-course guidance techniques for SLBMs and the more accurate navigation of submarines will steadily increase the accuracy of the missiles. SLBM warheads will eventually be fitted with terminal guidance, using radar, a laser, or some other device to guide them onto their targets after reentry into the earth's atmosphere. This could result in a CEP of a few tens of meters. SLBMs will then be so accurate as to cease being only countercity weapons and become nuclear war fighting weapons able to threaten even relatively small military targets.

Antisubmarine Warfare

A very large effort is being put into improving antisubmarine warfare (ASW) techniques by both the United States and the Soviet Union to detect and destroy enemy submarines, and these efforts will almost certainly lead, in time, to success. Even in the absence of a technological breakthrough—which cannot be discounted in spite of official reassurances—steady progress in limiting the damage that can be done by enemy strategic nuclear submarines will increase perceptions not only that a surprise attack may succeed but that it is essential.

In ASW, detection remains the critical element. Detection methods are being improved by increasing the sensitivity of sensors, improving the integration between various sensing systems, and better processing of data from sensors. Most current developments are, in other words, electronic.

All types of these ASW sensors are being improved—electromagnetic ones, based on radar, infrared sensors, lasers, and optics; acoustic ones, including active and passive sonar; and magnetic ones, in which the magnetic field disturbance caused by a submarine is measured. Airborne, spaceborne, ocean-surface, and sea-bottom sensors are being increasingly integrated and, therefore, made more effective. ASW aircraft, surface ships, and hunter-killer submarines are also being made increasingly complementary. Each system has special characteristics, and the integration of those that complement each other greatly enhances overall effectiveness.

The single most effective ASW weapon is the hunter-killer—a nuclear submarine equipped with sonar and other ASW sensors, underwater communications, a computer to analyze data from the sensors and to fire ASW weapons, such as torpedoes with active or passive acoustic terminal guidance and perhaps nuclear warheads. The United States and the Soviet Union each operate hunter-killer fleets, several dozen boats strong. A hunter-killer is designed to find and then follow an enemy strategic submarine until it is ordered to destroy it.

Cruise Missiles

Another strategic nuclear weapons system, the development of which depends almost entirely on microelectronics, is the modern cruise missile. This weapon is also extremely accurate and is clearly in the category of a nuclear war fighting weapon.

Cruise missiles are old weapons, dating back to the German V-1 or "buzz-bomb" of World War II, and it was soon after the war that the United States and Soviet Union began developing them. A variety of types were produced (surface-to-surface, surface-to-air, and air-to-surface) for both short-range (tactical) and long-range (strategic) uses.

In the early 1960s, U.S. long-range surface-to-surface cruise missiles were replaced by ballistic missiles. In 1972, the U.S. interest in cruise missiles revived, according to some, as a "bargaining chip" in the Strategic Arms Limitation Talks (SALT). A number of technological advances encouraged cruise missile development, the most important by far being the miniaturization of computers in terms of volume and weight for a given power output. The availability of an accurate data base describing the coordinates of potential targets was another important improvement. Very small but accurate missile guidance systems could thus be developed. For example, the McDonnell Douglas Terrain Contour Matching (TERCOM) system, which weighs only 37 grams, can guide a cruise missile to its target with a CEP of a few tens of meters. TERCOM uses an onboard computer to compare the terrain below the missile (scanned with a radar altimeter) with a preprogrammed flight path. Deviations from the planned flight path are corrected automatically. From maps made available using satellite mapping techniques, the positions of targets and contours of flight paths can be obtained with unprecedented accuracy. Targets could not be located accurately enough from earlier maps to make effective use of the cruise missile guidance systems. These guidance systems are essentially based on pattern recognition, one of the main military uses of artificial intelligence.

Using these new technologies, cruise missiles are being developed in the United States to be launched from air, sea, and ground platforms. Perhaps the most important characteristic of these cruise missiles is that the ratio of the payload carried to the physical weight of the missiles is relatively high (typically about 15 percent compared with a fraction of 1 percent for a typical ballistic missile).

The air-launched cruise missile (ALCM), to be carried and launched from strategic bombers, has, for example, a very small radar image and flies at subsonic speeds at very low altitudes (a couple of hundred meters over rough terrain and a few tens of meters over smooth ground). The missile, difficult to detect and destroy, makes defense against them equally difficult and costly, particularly if they are launched in large numbers. American strategic bombers are planned to carry up to twenty-four ALCMs equipped with nuclear warheads.

Airborne Warning and Control System (AWACS)

The AWACS is an exceedingly expensive piece of military high technology, incorporating a complicated system of microelectronics. The latest American AWACS aircraft, the E-3A, costs about $1.5 billion in research and development, and each aircraft costs about $100 million to produce, about six times the cost of the most sophisticated fighter.

According to current plans, a fleet of E-3As will eventually be built, eighteen of which scheduled to be bought by NATO. The aircraft will be used for the defense of North America, Central Europe, and the Greenland-United Kingdom gap. The Soviet Union also operates a fleet of AWACS aircraft, based on the Tu-126 Moss, but Soviet AWACS aircraft are much less sophisticated than the American version.

Many regard AWACS as a microelectronic white elephant in that the aircraft would be a prime and vulnerable target during war and likely to be lost soon after fighting begins. Since AWACS aircraft would have to operate in an environment of a great many electronic countermeasures (ECM), critics of the system claim that present-day radars are insufficiently effective against ECM. AWACS enthusiasts, however, argue that it can resist ECM better than current ground-based surveillance radars can, and that it would survive long enough in a hostile environment to fulfill a worldwide role. They emphasize the value of an early warning period, however short, in a crisis.

A modern AWACS aircraft, like the E-3A, not only performs the functions of an airborne early warning aircraft but, in addition, provides extensive command and control facilities for all friendly aircraft within its range—including interceptors, transport aircraft, reconnaissance aircraft, and so on. The identification and tracking of hostile aircraft and the control of friendly aircraft—at long range, at high and low altitudes, in all weathers, and over land or sea—are achieved by the use of highly sophisticated, beyond-the-horizon, look-down, high-pulse-frequency doppler surveillance radars, high-speed computers, and multipurpose display units.

The Automated Battlefield

In tactical warfare the most concentrated application of microelectronics is in the development of the automated battlefield. In the past fifteen years or so, considerable steps have been taken in automating warfare—in ground, sea, and air combat. This is perhaps the most dynamic of current military technologies, but it is also one shrouded in a great deal of secrecy. Most of the automatic systems being developed for the battlefield rely heavily on microelectronics. New electronic offensive weapons inevitably stimulate the development of electronic countermeasures against them. This, in turn, provokes the development of electronic counter-countermeasures, and so on.

Most battles have four distinct phases: enemy forces are located and identified; a decision is reached on how to deal with these forces; appropriate weapons are fired; and finally, damage to the enemy is assessed to find out if the sequence needs repeating.

In the automated battlefield, the enemy forces—men and vehicles—would be detected by remotely piloted vehicles or sensors planted on the ground. The data so collected would then be transmitted back to central computers that would decide on the action to be taken and then direct the weapons onto their targets. After the weapons had been fired, sensors on the battlefield and remotely piloted vehicles would assess the damage done, feed the information back to computers that would decide whether or not to fire more weapons, and so on.

Sensors. Sensors on the automated battlefield can be sensitive to light, sound, electromagnetic waves, infrared radiation, magnetic fields, pressure, and chemicals. We are most interested here, though, on sensors that transmit information about the enemy forces and movements by radio over long distances.

One of the main advantages of microelectronics in sensors is that although their size may be small, the power supply of their transmitters is strong enough to extend ranges to a few kilometers or more. As a rule, ground relay stations would, when necessary, be used to transmit sensor signals over very large distances to the central computers. Relay equipment may also be carried in high-flying aircraft, which may be remotely controlled.

Human operators inside control stations away from, or flying over, the battlefield would assess data from the sensors and direct appropriate weapons to the targets. But this phase of the battle may, in the future, also be automated. A computer could be programmed, for example, to react to information from sensors and issue orders for further reconnaissance or direct attack.

Stand-off Weapons. Guided weapons are usually designed for use on the automated battlefield—surface-to-surface missiles, air-to-surface missiles, and guided bombs. These may be fitted with automatic homing devices so that, once launched, the missiles will seek out the target without further external help.

A typical stand-off weapon contains an electro-optimal remote guidance and control system and a solid-propellant rocket engine giving the missile a range of about 100 kilometers. As the missile travels to its target area after launch from the parent aircraft, it returns data-link television pictures to an operator in the parent aircraft that has a display unit on which the terrain in sight of the TV camera mounted in the nose of the missile is described. In this way, the operator can change the missile's course, select a new target, and then lock the warhead onto the target by sending radio command signals to the missile over the data link.

These stand-off missiles will probably give way to a new generation of remotely piloted vehicles (RPVs) built for ground attack missions, with ranges of 200 kilometers or so. RPVs will also be used for reconnaissance, electronic warfare, and even air-to-air combat. The development of these small RPVs is made possible by microelectronic systems for guidance, control, and communications. Using TV cameras and data transmission links, and so on, an RPV could be controlled precisely from a safe distance from the battlefield, either by a pilot in a launch aircraft or by operators on the ground using TV pictures transmitted by the RPV.

A typical RPV currently under development will carry about 500 kilograms of cameras, electronic intelligence receivers, side-looking radar, and communications relays, will operate at altitudes of up to 25,000 meters with speeds of about 500 kilometers per hour, and will take off and land on a runway. For large-scale tactical warfare, a very large number of such RPVs would be used and controlled through computerized command and control centers. The vehicles are cheap enough to make such use in large numbers possible.

Modern Avionics

A modern combat aircraft is an excellent example of a weapons system whose performance relies on the most up-to-date electronics. The most advanced developments in avionics are to be found in air-superiority fighters, aircraft designed to seek, find, and destroy any type of enemy aircraft, whatever the weather. Such an aircraft may also be capable of other missions, such as air-to-surface attack.

A modern combat aircraft's avionics typically includes a lightweight radar system so that the aircraft can detect and track high-speed targets at great distances and at altitudes down to tree-top level. The tracking information is normally fed into the aircraft's central computer for the accurate launching of missiles or the firing of an internal gun. For close-range combat, the radar automatically projects the target on a head-up display that details all necessary information as symbols on a glass screen positioned at the pilot's eye level. The pilot can thus be automatically provided with the data required to intercept and destroy an enemy aircraft without taking his eyes off the target. The head-up display provides the pilot with navigation and control information under all conditions and with details about the aircraft's performance so that any faults that develop in any of the aircraft's systems can be detected as soon as they occur.

An "identification-friend-or-foe" system informs the pilot if an aircraft that has been detected by eye or radar is friend or enemy. And an air data computer and an attitude and heading reference set display information on the pitch, role, and magnetic heading of the aircraft. This, together with an inertial navigation set, enables the pilot to navigate anywhere in the world.

The United States is now considering a so-called quick-reaction interceptor aircraft, with speeds greater than Mach 3 (three times the speed of sound) and extremely rapid acceleration, for operations with future air-defense weapons systems. Such an aircraft may have air-superiority and tactical-strike roles in addition to its role as interceptor. Even faster aircraft with hypersonic (up to Mach 10) speeds are also under active consideration. Advanced avionics are making such extraordinary combat aircraft feasible.

A typical air-superiority fighter would be armed with air-to-air missiles of short and medium ranges. The U.S. Sparrow is an example of an all-weather, all-aspect, air-to-air missile. It uses a continuous wave, semiactive radar guidance system and carries a proximity-fused, high-explosive warhead weighting about 5 kilograms. A solid-propellant rocket propels the missile at a speed of Mach 3 over a range of about 50 kilometers. The very latest Sparrow air-to-air missiles combine radar and infrared in their target-seeking system. Missiles with semiactive radar guidance require the target to be continuously illuminated. The aircraft's radar transmits continuous-wave, illuminating signals so that, after launch, the missile can home in on the reflections of these signals from the target. Because the aircraft's radar antenna must be directed at the target until the missile hits it, only one target can be engaged at a time. But heat-seeking missiles carry their own terminal guidance and, therefore, need only be launched in the direction of the target.

ARMS RACE BETWEEN THE SUPERPOWERS

Microelectronics has become the main factor in the technological arms race between the two superpowers. By continuous effort in military technological development, the Americans are striving to retain a technological superiority in as many weapons systems as possible, a superiority the Soviets are seeking to eliminate. A fundamental American belief is that, provided sufficient resources are devoted to research, the nature of their political system is such as to encourage innovation more than other political systems do. This then is a field of activity, and possibly the only one, in which the United States hopes to be able to keep permanently ahead of the rest of the world. Failure would, so it is believed, bring military or political disaster. But such opinions, based on perceptions rather than facts, are by no means confined to Americans. Political leaders in general appear to believe that military technological superiority prevents potential adversaries from applying pressures and blackmail in international affairs, or that this superiority brings victory in war.

The struggle by both the United States and the Soviet Union for military preeminence is by far the most important single cause of the ongoing arms race (nuclear and conventional) between them. The arms race will undoubtedly continue until both sides cease attaching such enormous political importance to gaining and retaining technological superiority. Past arms races have led to wars or to economic collapse. There is no reason to believe that the modern arms race is any exception.

Although the most dramatic advances in weaponry have occurred in strategic arms, continual improvements are being made in the quality of virtually all tactical weapons. In very many cases, the advances come from microelectronic developments.

The dynamics of superpower military technology can be illustrated by one example, the military use of high-energy lasers, though many others could be given. The rapid progress in production of high-energy lasers (continuous-wave power outputs of a few hundred kilowatts can now be achieved) has stimulated great interest in the development of thermal laser weapons. Likely applications for the first generation of these revolutionary weapons, in combination with appropriate electronics, include ground-based air defense against low-flying aircraft, missiles and remotely piloted vehicles, and air-to-air combat. The range of military applications of high-energy lasers will undoubtedly increase as their power output increases, including perhaps, the development of a space-based defense system against enemy ballistic missile warheads.

MILITARY SCIENCE AND THE INCREASING PROBABILITY OF NUCLEAR WORLD WAR

Even though the ever increasing gulf between the military power of the United States and the Soviet Union as compared with all other countries has serious consequences for world security, it is overshadowed by the increasing probability of nuclear world war brought about by developments within the superpowers, mainly military technological developments.

The world's arsenals contain tens of thousands of nuclear weapons, probably topping 60,000. The total explosive power of these weapons is equivalent to about 1.25 million Hiroshima bombs, or about 4 tons of TNT for every man, woman, and child on earth. If all, or a significant portion of them, were used, the consequences would be beyond imagination.

All the major cities in the Northern Hemisphere, where most nuclear warheads are aimed, would be destroyed (on average, *each* is targeted by the equivalent of 2,000 Hiroshima bombs). Most of the urban population would be killed by blast and fire, the rural population by radiation from fallout. Many millions of people in the Southern Hemisphere would be killed by radiation. And the disaster would not end even there. The unpredictable (and, therefore, normally ignored) long-term effects might well include changes in the global climate, severe genetic damage, and depletion of the ozone layer that protects life on earth from excessive ultraviolet radiation. No scientists can convincingly assure us that human life would survive a nuclear world war.

Utterly catastrophic though nuclear world war might be, its probability is steadily increasing because military scientists are developing weapons that will be seen as suitable for *fighting* rather than *deterring* a nuclear war. As we have seen, these new weapons include very accurate and reliable ballistic missiles with warheads that can be aimed at smaller and, therefore, a greater number of military targets than in the past. In other words, the day is coming when one country might hope to destroy its enemy's nuclear retaliatory capability by striking first.

In this context, a first strike does not mean the ability to destroy totally the other side's strategic nuclear forces in a surprise attack. What it does mean is that the attacking power *perceives* that it can destroy enough of the enemy's retaliatory forces in a surprise attack to reduce the casualties it receives in a retaliatory strike to a number regarded as "acceptable" for a given political goal. (In nuclear issues, as in almost all matters of international politics, perceptions determine events; facts do not.) The more reckless the political leadership is and the more tense superpower relations are, the greater this number of casualties is likely to be. And in its calculations the attacker is likely to make assumptions about the performance of its own and its enemy's nuclear forces to suit its arguments. Specifically, military calculations are likely to be based mainly on estimates of prompt deaths and ignore the uncertain long-term effects of a nuclear war, which may well be far more lethal than the early effects. In times of crisis, political leaders tend to listen to their military chiefs rather than their scientific advisers.

The direct attack by one superpower on the other may not be the most likely way in which a nuclear war will start. A more probable way may be the escalation of a future conflict in a Third World region. What may begin as a conventional war escalates to a nuclear war using the nuclear weapons of the local powers. This situation could develop, in turn, into an all-out nuclear war between the superpowers.

This escalation is most likely to occur if the superpowers are already involved in the Third World conflict, because they had supplied the conventional weapons for the original fighting. Modern war uses munitions, particularly missiles, so rapidly that Third World countries at war are in continuous need of new supplies of weapons. In this way, the arms supplier virtually guarantees the survival of its client at war. It is because of its contribution to the threat of a third world war that the international arms trade is so dangerous.

The superpowers use these conflicts to test their most sophisticated weapons. Massive use of the most advanced microelectronics in modern weapons has exerted much influence on this situation: it increases the demand for sophisticated weapons by Third World countries so that they can have the most up-to-date—and, it is assumed, the most effective—weapons; it increases the temptation of the superpowers to test continually the electronics in their missiles and other weapons by their use in Third World conflicts; it increases the violence of wars; and it increases the probability of an escalation of conflicts.

The superpowers can be blamed for the danger of nuclear catastrophe that we all face. They, and their main allies, after all, spend the bulk—about 80 percent—of world military expenditure and own the vast majority of nuclear warheads in the world today. The Third World, meanwhile, spends relatively little—about 13 percent—of total world military expenditure. The superpowers also supply most—about 75 percent—of the arms transferred to the Third World. Nevertheless, most conflicts take place in the Third World, and there have been about 140 such conflicts since World War II. And, as already noted, any future Third World conflict may escalate to nuclear world war.

Improvements in warhead design and missile accuracy, which have virtually reached their theoretical limits, are just two examples indicating the incredible progress made by military technology since World War II, and in the next thirty years, we can expect even more military technological revolutions. Many of these developments—which more often than not will be based on advances in microelectronics—will contribute to perceptions that a nuclear war is fightable and winnable, and that a first strike is feasible and even essential, on the argument that unless we strike now the other side will soon do so.

An excellent example of this sort of thinking can be seen in the development of antisubmarine warfare. Now that land-based intercontinental ballistic missiles are, or soon will be, vulnerable to a first strike by missiles, nuclear deterrence depends on the invulnerability of submarine-based ballistic missiles. The fact that the superpowers continue working so energetically on antisubmarine warfare shows that they are unable to control military science and technology even though this activity is jeopardizing their policy of nuclear deterrence, a policy the leaderships desperately want to maintain.

Perhaps one should emphasize here the military use of space. The application of microelectronics is the main reason for, and makes possible, the rapidly increasing militarization of space. The development and deployment of space-based, first-strike technologies is particularly disturbing. These include, in addition to reconnaissance satellites ("spies in space"), early warning systems, navigational systems, command, control, communication, and intelligence systems, antisubmarine warfare systems, and ballistic missile defense systems. Since the space age began in 1957, about 1,200 satellites have been launched. About 75 percent of them have been for military purposes.

The large group of scientists who rely entirely on military money for support is, of course, a powerful political lobby. Moreover, vast bureaucracies have grown up inside the great powers to deal with military matters. (As many civilians are paid out of military budgets as there are troops in uniform.) Academics and bureaucrats join with the military and defense industries to form an academic-bureaucratic, military-industrial complex intent on maintaining and increasing military budgets and arguing for the use of every conceivable technological advance for military purposes. The complex has so much political power as to be almost politically irresistible. In fact, the nuclear arms race is now totally out of the control of political leaders.

And this is as true in the Soviet Union as it is in the United States. The uncontrolled nuclear arms race is without doubt the single greatest threat to our survival.

This is not to deny that great efforts have been made to control military technology and to stop the nuclear arms race between the Americans and the Soviets. Since World War II many of the world's most brilliant individuals have been actively involved in these efforts. No other problem has received so much attention in the United Nations and other international forums. Yet, because of the enormous political influence of those groups that continually press for greater military efforts, nuclear and other arms races go on just as fast as human skill in the American and Soviet societies allows. We are being driven toward nuclear world war by the sheer momentum of military technology.

4 THE GENESIS OF NUCLEAR POWER

Gordon Thompson

After heady days in the early 1970s, the prospects for the nuclear power industry have grown consistently poorer. In 1974, the U.S. Atomic Energy Commission (AEC) projected that nuclear plants would generate 41 percent of U.S. electricity in 1985, rising to 62 percent at the end of the century.[1] By 1983, however, nuclear generation accounted for only 13 percent of all U.S. electrical generation, a percentage that has changed little since 1977. Many nuclear plants under construction or on order have been cancelled. The total capacity of actual or planned reactors in the United States reached a peak of 236,000 megawatts (MW) in 1975, declining to 134,000 MW in 1983.[2] More cancellations are likely. As an example, the Marble Hill plant in Indiana was cancelled in January 1984, despite the expenditure of $2.5 billion on its construction.[3]

With this decline in the nuclear industry's fortunes has come a collapse of plans for "closing the nuclear fuel cycle," using reprocessing plants and breeder reactors. From the early days of the industry, it was assumed that plutonium would be separated from the spent fuel discharged by conventional reactors in an operation that would be performed in reprocessing plants. The separated plutonium would then be used as fuel in breeder reactors, which would, in addition to generating electricity, convert nonfissile uranium (U-238, which represents 99 percent of natural uranium) to a quantity of plutonium slightly greater than the quantity the breeder reactor consumed. In this manner, the life of uranium reserves would be greatly extended.

Despite the support of the Reagan administration, fundamentally poor economics have blocked plans to demonstrate this plutonium cycle. The Barnwell reprocessing plant in South Carolina has failed to attract private funding sufficient for its completion and will probably be dismantled. Congress, after a prolonged struggle, cancelled Tennessee's Clinch River Breeder Reactor project in 1983 when it became evident that a significant private contribution to its escalating construction budget would not be forthcoming.

The U.S. experience has been repeated in most countries with nuclear power programs. Only a few countries, such as France and Japan within the industrialized countries, and South Korea and Taiwan in the developing world, are proceeding with vigorous nuclear power programs. Even these countries, though, are revising their expectations downward in the light of adverse trends in nuclear economics.[4]

Nuclear power's failure to thrive in the energy market arises from a variety of factors, the major causes being declining growth rates for electricity demand and rising construction costs for nuclear plants. Electricity demand grew rapidly after World War II, with U.S. utilities typically seeing growth of 7 percent per annum (which represents a doubling of electricity demand every 10 years). Since the early 1970s, however, rising energy prices have curbed growth in demand for all energy forms, including electricity. In consequence, programs for construction of new generating plants have been scaled back. Nuclear plants have been particularly affected because escalation of their construction costs has made them uncompetitive with alternatives, especially coal plants.

Construction costs for nuclear plants have escalated above those for coal plants because difficulties have arisen in the pursuit of safety and reliability. Reactor suppliers, architect-engineers, and utilities have struggled to compensate for operating failures and new safety regulations. As reactor operating experience has accrued, new problems have continued to emerge. Partly under mandate from the U.S. Nuclear Regulatory Commission (NRC), and partly on their own initiative, utilities have made numerous modifications to nuclear plants under construction and in operation.

In this manner, the nuclear industry is suffering the legacy of its pioneers, who launched the industry with inflated expectations and without regard for the adverse implications that have since emerged. The seeds of that legacy lie, to a considerable degree, in the military origins of the industry.

THE NAVY AND NUCLEAR POWER

H. G. Wells, in his 1913 science-fiction novel *The World Set Free*, saw a future of depleted natural resources. Fossil fuels, metal ores, forests, and water supplies all suffered from reckless exploitation. However, Wells saw a solution to this problem in the harnessing of nuclear energy, the discovery of which he placed in the year 1933. By the 1950s, he saw the whole world running on cheap and plentiful nuclear energy, " . . . once more Eden."[5]

As we now know, Wells was remarkably prescient about the liberation of nuclear energy, but overly optimistic about its application. While nuclear energy has come to play a relatively small role in our energy supply, its military application has proceeded to the point where it could extinguish our civilization and, if trends continue, our species. Recent research has shown that a nuclear war could lead to massive ecological disruption over much of the Earth's land surface.[6,7,8]

Immediately after nuclear weapons were used at Hiroshima and Nagasaki, there was a flood of predictions about the beneficial uses of the new energy form. Robert Hutchins, chancellor of the University of Chicago, predicted that " . . . a very few individuals working a few hours a day at very easy tasks in the central atomic power plant will provide all the heat, light and power required by the community and these utilities will be so cheap that their cost can hardly be reckoned." David Lilienthal, soon to be the first head of the AEC, wrote of the "almost limitless beneficial applications of atomic energy."[9]

While the wartime Manhattan Project had been under the control of the army and had a strictly military purpose, the AEC, formed in August 1946, was charged with developing both civil and military technologies. This institutional arrangement expressed a widely felt desire that the new knowledge, still largely secret because of its weapons implications, should be used for constructive purposes. The dual nature of nuclear technology was captured by President Truman, in a remark to AEC Chairman Lilienthal in April 1947: "Come in to see me any time, just any time. I'll always be glad to see you. You have the most important thing there is. You must make a blessing of it or [and a half-grin as he pointed to a large globe in the corner of his office] we'll blow that all to smithereens."[10]

Perhaps as a reaction to the destructive potential of the atomic age, perhaps from natural optimism, the proponents of commercial nuclear power pressed forward despite skepticism in influential circles. In 1947, the AEC's General Advisory Committee, chaired by Robert Oppenheimer, reported that it saw numerous technical and economic problems clouding the future of nuclear power. Moreover, this committee felt that the AEC's reactor research program was bureaucratic and uninspired and lacked first-rate scientists and engineers.

A variety of reactor types were, nonetheless, developed in the national laboratories. In 1951, the world's first nuclear-generated electricity was produced by the Experimental Breeder Reactor I in Idaho. It is ironic, given the commercial failure of liquid-metal-cooled breeder reactors over the last three decades, that the first nuclear electricity should have come from such a reactor.

Without a doubt, however, the major influence on reactor development in this period was Captain (later Admiral) Hyman Rickover. As head of the Electrical Section of the Bureau of Ships during World War II, Rickover gained a reputation for ruthless energy and efficiency. In 1946, Rickover, with a delegation of other naval officers, attended a training course on nuclear technology at the Clinton Laboratory (later the Oak Ridge National Laboratory). From Alvin Weinberg, director of this laboratory, Rickover picked up the idea of a pressurized-water reactor (PWR). This reactor, by virtue of its relative simplicity and compact design, proved to be well-suited to submarine propulsion.

In 1948, Rickover set himself the goal of having a nuclear-powered submarine operational by January 1955. For this purpose, he obtained an unusual joint appointment: head of the AEC's Naval Reactors Branch and head of the navy's equivalent office. Also, he cultivated links with Congress, especially the powerful Joint Committee on Atomic Energy. Finally, he worked closely with private industry, particularly Westinghouse Corporation. By cooperating with Rickover, industry gained research experience at public expense. In return, Rickover obtained the technology he wanted, on his terms. Under this arrangement, Westinghouse set up a new laboratory (the Bettis Laboratory) near Pittsburgh, with AEC funding. Rickover then exerted almost total control over work at Bettis. The deadline was met: the submarine *Nautilus* cast off from the pier of the Electric Boat Company on January 17, 1955. A Westinghouse-developed PWR provided the motive power.

The General Electric Corporation (GE) also flirted with Rickover but more cautiously than Westinghouse. After the war, GE agreed to operate the plutonium-producing reactors at Hanford, on condition that the AEC would fund a nuclear research laboratory for the company. This laboratory was built near GE's headquarters at Schenectady, New York. In 1950, GE began to work for Rickover and built a sodium-cooled submarine reactor. This reactor proved, however, to be uncompetitive with the PWR.

From the Argonne National Laboratory, GE picked up the concept of a boiling-water reactor (BWR). GE went on to develop this concept at its own expense, creating for this purpose a new laboratory at San Jose, California. Thus were born the two reactor types that have come to dominate the world's nuclear industry, the PWR and the BWR. As of mid-1982, PWRs accounted for 56 percent of the capacity of all commercial reactors operating in the world, and 66 percent of the capacity of those under construction. BWRs accounted for 26 percent of the capacity of operating reactors, and 21 percent of the capacity of reactors under construction.[11]

Under pressure from the Joint Committee on Atomic Energy, Congress in 1953 allocated funds to the AEC for construction of a pilot nuclear power plant. It was felt that the United States must compete with nuclear power programs in the USSR and in Britain. Given Rickover's preeminent role at this time, it was natural that he should be entrusted with the task and that a PWR should be used. In a cooperative venture with the Duquesne Light Company, the AEC built a 60 MW Westinghouse PWR at Shippingport, Pennsylvania. The project used technology developed for a nuclear-powered aircraft carrier that was cancelled early in 1953.

In December 1953, President Eisenhower made his famous "Atoms for Peace" speech to the UN General Assembly, a speech that sought to redirect competition in nuclear arms toward efforts to develop peaceful nuclear technology. It was thus easy for Rickover to persuade Eisenhower to participate in a theatrical ground-breaking ceremony for the Shippingport project in September 1954. While appearing on television in Denver, Colorado, Eisenhower waved a radioactive "wand" that activated an unmanned bulldozer in Shippingport, breaking the first ground for the plant.

Construction of the Shippingport plant proceeded smoothly. The reactor achieved its design power in December 1957 and continued operating in various roles until finally shut down in 1982.

The Atomic Energy Act of 1954 facilitated the involvement of private industry in nuclear power development and represented a view in Congress and the AEC that the time had come for commercialization of the new technology. In 1955, the Power Reactor Demonstration Program was initiated by the AEC. This program, which continued in various forms until 1962, was intended to attract private industry into this new field. A variety of subsidies and technology transfer measures were employed.[12]

As stated earlier, a variety of reactor types were developed in the national laboratories. Thus, the nuclear power plant projects encouraged by the Power Reactor Demonstration Program were not restricted to PWRs and BWRs. The commercial nuclear plants ordered through 1962 included a sodium-cooled fast-breeder reactor (FBR), a sodium-cooled graphite-moderated reactor (SGR), a high-temperature gas-cooled reactor (HTGR), a heavy-water-moderated reactor (HWR), and an organic-liquid-moderated reactor (OMR). Some characteristics and operating histories of these early plants are shown in Table 4-1.

As suggested by this table, and confirmed from later experience, the alternative reactor types did not take root in the nuclear energy field. The final entry in Table 4-1 (Haddam Neck) sets the tone for later developments—this plant is a scaled-up (580 MW) Westinghouse PWR. It was soon followed by the similarly scaled-up (620 MW) Nine Mile Point BWR—a GE reactor that was ordered in 1963 and first generated electricity in 1969. Over the period from 1962 to the present, only two power reactors other than PWRs and BWRs have been ordered and built in the United States—the 850 MW N-Reactor in Hanford, Washington (whose primary function was to make plutonium for nuclear weapons), and the 330 MW Fort St. Vrain HTGR in Colorado. Westinghouse has maintained its early lead in nuclear power technology—its PWRs accounted for 44 percent of the capacity of reactors operating or under construction in the United States in 1983. General Electric BWRs accounted for 32 percent of that capacity, with Combustion Engineering and Babcock-Wilcox PWRs accounting for 13 percent and 10 percent, respectively.[13]

Table 4-1. Characteristics and Operating Histories of U.S. Commercial Nuclear Power Plants Ordered through 1962.

Plant	Type	Capacity (MW)	Year of: Order	Initial Operation	Final Shutdown
Shippingport	PWR	60	1953	1957	1982
Indian Point	PWR	265	1955	1962	1974
Dresden	BWR	200	1955	1960	–
Yankee	PWR	175	1956	1960	–
Fermi	FBR	61	1957	1966	1972
Pathfinder	BWR	59	1957	1966	1967
Hallam	SGR	75	1957	1963	1964
Humboldt Bay	BWR	65	1958	1963	1976
Elk River	BWR	22	1958	1963	1968
Peach Bottom	HTGR	40	1958	1967	1974
Carolinas-Virginia	HWR	17	1959	1963	1967
Piqua	OMR	11	1959	1963	1966
Big Rock Point	BWR	72	1959	1962	–
BONUS	BWR	17	1960	1964	1968
LaCrosse	BWR	50	1962	1968	–
Haddam Neck	PWR	580	1962	1967	–

Source: U.S. Department of Energy, *U.S. Central Station Nuclear Electric Generating Units: Significant Milestones* (October 1983).

Alternative reactor types were also developed in other countries. Both Britain and France commenced their nuclear programs using gas-cooled graphite-moderated reactors (GCRs). The initial versions were designed primarily for production of plutonium for nuclear weapons—later versions were primarily intended for electricity production. In the late 1960s, the British began a second generation of this reactor type, the advanced gas-cooled reactor (AGR). The French, however, switched to PWRs (of Westinghouse design). More recently, Britain's Central Electricity Generating Board has also turned to the PWR (again, using Westinghouse technology). A public inquiry is being held during the years 1983 and 1984 to advise the British government on the wisdom of this choice.

The Soviets adopted the water-cooled graphite-moderated reactor (LWGR) for their early nuclear plants but gradually switched to

PWRs. As of mid-1982, LWGRs accounted for 64 percent of the capacity of power reactors operating in the Soviet Union, PWRs accounting for 31 percent. However, LWGRs accounted for only 35 percent of the capacity of reactors under construction, whereas PWRs accounted for 65 percent.[14] It may be assumed that the Soviet PWR was developed from submarine reactors, as in the United States. The LWGR appears to have its origins in the production of plutonium for nuclear weapons, as in France and Britain.

Alone among the nonnuclear-weapon countries, Canada developed an indigenous nuclear power technology. This resulted in a generation of CANDU reactors, moderated and cooled by heavy water and similar to the Carolinas-Virginia prototype reactor built in the United States (see Table 4-1).

Canada's CANDU reactors, and the GCRs of Britain and France, are notable in that they are fuelled by natural uranium. By contrast, PWRs and BWRs are fuelled with uranium in which the content of the fissile isotope U-235 has been enriched from its natural level (0.7 percent) to about 3 percent. Thus, the early reactor choices made by these three countries reflected their lack of uranium enrichment capacity. As is evident from the current predominance of PWRs and BWRs, there is now, however, a worldwide abundance of enrichment capacity.

The United States built three huge enrichment plants (in Tennessee, Kentucky, and Ohio) in the 1940s and 1950s. Their primary purpose was to supply highly enriched (more than 90 percent U-235) uranium for nuclear weapons (a small amount of highly enriched uranium has also been used to fuel naval propulsion reactors). Since the mid-1960s, as the number of nuclear weapons has stabilized, and plutonium has become the preferred weapons material, this enrichment capacity has become available for the production of low-enriched uranium as a fuel for nuclear power plants. Other countries have also become suppliers of enrichment services.

The military origins of nuclear power are reflected in the expenditures that have nurtured it, particularly in the early years. Table 4-2 shows research and development (R&D) expenditures by the U.S. government up to 1962, when the industry began to operate in a more commercial mode. It is clear that R&D on naval reactors exerted a dominant influence well into the 1950s. The amounts involved are small by current standards, even when corrected for infla-

Table 4-2. Expenditures by the U.S. Government on Research and Development of Commercial Nuclear Power, 1950-1962 (*millions of current dollars*).

Year	*Direct Civilian Expenditures*[a]	*Additional Contribution from Military Expenditures*[b]
1950	3.1	5.1
1951	5.1	17.7
1952	6.3	24.5
1953	10.1	38.8
1954	18.9	35.5
1955	29.9	32.8
1956	55.1	51.4
1957	97.5	99.0
1958	148.1	103.6
1959	184.7	96.5
1960	243.8	79.3
1961	282.2	82.3
1962	271.8	47.7

a. Includes work on breeder reactors but not on fusion.

b. Includes 100 percent of naval reactor expenditures through 1955; 50 percent thereafter.

Source: Energy Information Administration, U.S. Department of Energy, *Federal Support for Nuclear Power: Reactor Design and the Fuel Cycle*, Energy Policy Study, vol. 13, AR/EU/81-01 (February 1981).

tion. In part, this reflects a more frugal age, but it also shows the relative simplicity of the naval and early commercial reactors. Current problems in reactor safety owe much to an ambitious scaling-up, from 60 MW for the Shippingport reactor to over 1,000 MW for current plants.

Subsidies did not cease when the AEC's early commercialization effort (the Power Reactor Demonstration Program) was over. Industry benefited from continuing government-sponsored R&D on reactors and on other aspects of the nuclear fuel cycle, and from access to cheap uranium and enrichment services. It has been estimated that total subsidies to the nuclear power industry through 1981 amounted to $49 billion in 1980 dollars, implying that the true cost of nuclear-generated electricity at the close of the 1970s was higher than the cost experienced by utilities by a factor of 50-100 percent.[15]

Although these cumulative subsidies dwarf the amounts shown in Table 4-2, this does not alter the importance of the early military influences. Rickover's work on nuclear submarines, and the availability of uranium enrichment capacity in the United States, set the pattern that the nuclear industry has ultimately followed worldwide. It seems that with a new industry, as with a child, early influences are crucial to later development.

CONCLUSIONS

Although the PWR/BWR combination has swept aside competing reactor types, there is growing concern that it is leading the nuclear power industry into a dead-end. Recurring safety problems, together with concerns over weapons proliferation and the management of radioactive waste, are compounding an already adverse economic environment. Thus, a number of institutions and individuals, including pioneers of the industry, are seeking a new path.

Former AEC Chairman David Lilienthal has written:

> We have poured countless millions into the light water method [i.e., PWRs and BWRs]. Its apparent economic advantages lured us into rushing into it, as though it were the best feasible answer, and downgrading with scant attention the other possibilities. Now it is time to recognize as a matter of national policy that the method is not good enough, not safe enough, not the right answer.[16]

Alvin Weinberg, inventor of the PWR, now head of the Institute for Energy Analysis (located in Oak Ridge, Tennessee), shares Lilienthal's concern. His institute has embarked on a study of the Second Nuclear Era, an effort intended to chart a path for nuclear power that avoids the pitfalls of the First Era. One of the elements of the study is a search for a "super-safe" reactor—a search that has focussed on the Process Inherent Ultimate Safe (PIUS) reactor concept recently developed by Sweden's ASEA-Atom company. It is ironic that this concept employs pressurized light water as a coolant and moderator, just as in the traditional PWR. However, the reactor is submerged in a large pool of water held in a large concrete pressure vessel located below ground level. This configuration is very different from the compact PWR that emerged from submarine practice.[17]

It remains to be seen if an economically competitive "super-safe" reactor can be developed. Even if it were developed, however, serious problems would remain. Diffusion of nuclear power technology has increased the likelihood of nuclear weapons proliferation, a trend that will be compounded if plutonium separation is widely practiced. Final disposal of radioactive wastes remains an intractable problem, both technically and politically. Moreover, commercial nuclear power is having difficulty shedding its military past.

In 1981, the U.S. Secretary of Energy openly proposed the extraction of plutonium from spent fuel generated by U.S. commercial reactors, in order to meet the plutonium needs of an expanded program of nuclear weapons production.[18] This proposal was subsequently curbed by a vote in the U.S. Senate. More recently, the French government has refused to rule out the use of the Superphenix breeder reactor to produce plutonium for French nuclear weapons.[19] Public concern about the nuclear arms race thus remains justifiably linked with concern about nuclear power.

To summarize, nuclear power began as an offshoot of military activities and continues to be influenced by military needs. This connection, initially helpful, has played a major role in bringing the industry to its present predicament. Had the industry's origins been more commercial, technical optimism would have been more effectively curbed. With hindsight, it is clear that a good deal of the optimism reflected a sincere desire to find a peaceful application for nuclear energy. To some degree also, "Atoms for Peace" provided a convenient cloak for a vast program of nuclear armament that was subject to relatively little public debate.

Our nuclear power path has not been followed without costs. Other energy options have been foreclosed. It is noteworthy that a special commission on natural resources, chaired by William Paley, which reported to President Eisenhower in 1953, said of future energy supply: "Nuclear fuels, for various technical reasons, are unlikely ever to bear more than about one-fifth the load. . . . It is time for aggressive research in the whole field of solar energy—an effort in which the United States could make an immense contribution to the welfare of the world."[20]

After thirty years, during which expenditures for solar energy development have been a tiny fraction of those devoted to nuclear power, we are little further forward in solving our long-term energy-supply problems. When making further commitments, we should be careful to learn from this experience.

NOTES TO CHAPTER 4

1. U.S. Atomic Energy Commission, *Environmental Statement: Liquid Metal Fast Breeder Reactor Program*, WASH-1535 (December 1974).
2. Energy Information Administration, U.S. Department of Energy, *Monthly Energy Review* (November 1983).
3. J. Rangel, *The New York Times*, January 17, 1984, page A1.
4. C. Flavin, *Nuclear Power: The Market Test*, Worldwatch Paper 57, Worldwatch Institute (Washington, D.C., December 1983).
5. Quoted in: A. E. Weiss, *The Nuclear Question* (New York: Harcourt Brace Jovanovich, 1981), Ch. 1.
6. Royal Swedish Academy of Sciences, *Ambio* XI, no. 2–3 (1982).
7. R. P. Turco, et al., "Nuclear Winter: Global Consequences of Multiple Nuclear Explosions," *Science* (December 23, 1983): 1,283–1,292.
8. P. R. Ehrlich, et al., "Long-Term Biological Consequences of Nuclear War," *Science* (December 23, 1983): 1,293–1,300.
9. Quoted in: D. Ford, *Cult of the Atom*: The Secret Papers of the Atomic Energy Commission (New York: Simon and Schuster, 1982), Part 1.
10. D. E. Lilienthal, *Atomic Energy: A New Start* (New York: Harper & Row, 1980), p. 2.
11. International Atomic Energy Agency, *Nuclear Power Reactors in the World* (Vienna, September 1982).
12. Energy Information Administration, U.S. Department of Energy, *Federal Support for Nuclear Power: Reactor Design and the Fuel Cycle*, Energy Policy Study, vol. 13, AR/EU/81-01 (February, 1981).
13. U.S. Department of Energy, *U.S. Central Station Nuclear Electric Generating Units: Significant Milestones* (October 1983).
14. International Atomic Energy Agency, *Nuclear Power Reactors.*
15. J. Bowring (National Center for Economic Alternatives), "Federal Subsidies to Nuclear Power: Reactor Design and the Fuel Cycle," prepared testimony for hearings on *Nuclear Fuel Cycle Policy and the Future of Nuclear Power*, Subcommittee on Oversight and Investigations, House Interior Committee, U.S. Congress, October 23, 1981.
16. Lilienthal, *Atomic Energy: A New Start*, p. 17.
17. I. Spiewak, "The Second Nuclear Era," *Institute for Energy Analysis Newsletter* (Summer 1983).
18. James Edwards, U.S. Secretary of Energy, address to the Energy Research Advisory Board, U.S. Department of Energy, September 3, 1981.
19. A. MacLachlan, "France Not Ruling Out Using Superphenix Plutonium for Weapons," *Nucleonics Week* (April 28, 1983): 2–4.
20. Quoted in: P. Pringle and J. Spigelman, *The Nuclear Barons* (New York: Holt, Rinehart and Winston, 1981), p. 164.

REFERENCES

Bupp, Irvin C., and Jean-Claude Derian. *Light Water: How the Nuclear Dream Dissolved.* New York: Basic Books, 1978.

Ford, Daniel, *Cult of the Atom: The Secret Papers of the Atomic Energy Commission.* New York: Simon and Schuster, 1982.

Lilienthal, David E. *Atomic Energy: A New Start.* New York: Harper & Row, 1980.

Pringle, Peter, and James Spigelman. *The Nuclear Barons.* New York: Holt, Rinehart and Winston, 1981.

5 THE MILITARY AND SEMICONDUCTORS

Robert DeGrasse

Government officials, economists, and scientists have often claimed that military spending encourages technological progress and results in civilian "spinoffs." Technologies that are cited as having profited from military spending include: aerodynamics, jet engines, computers, electronics, numerically-controlled machine tools, lasers, and nuclear power. Two broad arguments have been advanced to explain how military efforts enhance technology. In one, military demands are seen as a prod that continually encourages scientists and engineers to expand the frontiers of knowledge. By setting higher performance standards than are typically encountered, military projects are said to increase the "state of the art."[1] As a senior economist in the Pentagon argues:

> Defense sets goals that are difficult to meet; and our new programs often tax the limits of technology. Only the Department of Defense's budget is rich enough to experiment with new approaches to complex problems. It is my belief that we cannot foretell exactly the future path that technology must take in the quest for new commercial applications and solutions to non-defense problems. In the same sense that we seed the clouds in the hope for rain, so too we seed our research laboratories in the hope for finding solutions to difficult problems.[2]

A second argument, viewing military spending as a source of demand for new products, typically runs this way: "By providing an initial market and premium prices for major advances, defense purchasers speeded their introduction into use."[3] Transistors and integrated circuits are good examples of innovations that benefited from defense purchases when the price was significantly higher than civilian firms were willing to pay. Purchases of these goods for defense and space applications allowed manufacturers to improve their products and reduce costs by gaining production experience, a phenomenon known in the field as "coming down the learning curve."

The military's substantial funding for advanced weapons systems and research and development has certainly yielded some benefits. To be seen in perspective, however, positive effects must be weighed against any negative influences arms programs may have on technological advancement. At least three broad areas should be considered. First, military-oriented research and production diverts scientists and engineers from civilian pursuits. As a result, we are left with fewer people to develop civilian technologies such as consumer electronics, fuel-efficient cars, alternative energy systems, and mass transit. This drawback is particularly worrisome when high technology resources are limited, as they are today. Competition between the Pentagon and private industry for highly skilled labor, key subcomponents, and raw materials can drive up the price of American high technology products, making them less competitive in the world market.

Second, military-oriented programs can distort a new technology by encouraging applications that are too sophisticated to be marketed commercially. British and French experience with the SuperSonic Transport (SST) program is one example of this problem. While the United States wisely chose not to develop a civilian SST, our European allies proved that the military's pioneering research on flying at supersonic speeds did not have widespread commercial application. Nuclear power, with its unsolved safety problems and excessive cost, is another example. As we shall see later, military-sponsored programs designed to spur development of faster integrated circuits and more automated machine tools may also distort their development for commercial markets.

Finally, at the political level, we must assess the implications of according the military significant control over science and technology policy. While many of our politicians, including President Reagan, extoll the virtues of the free market, they still allow the Pentagon to control about a third of all public and private research and development funds and to purchase over 10 percent of the durable manufactured goods produced in our economy. These expenditures influence our technological and economic direction just as Japanese government policies controlling trade, encouraging investment, and subsidizing research influence that nation's direction. Japan's goal is economic growth, whereas our government largely aims for technological superiority in armaments.

In this chapter, we first use broad economic data to assess the technological costs and benefits of military expenditures. The key question is: What resources have been devoted to military technology and how has their diversion affected overall technological progress? We will attempt to answer that question in depth by examining the role that military programs have played in the development of the electronics industry.

MEASURING THE MILITARY'S IMPACT ON TECHNOLOGY

Between 1960 and 1973, Defense Department contracts for hardware averaged 16.9 percent of the durable manufactured goods sold in the United States. Since then, hardware contracts have averaged 10.9 percent of durable manufactured goods production. Of the major hard goods purchased by the Pentagon over the past three decades, at least 70 percent have been components of high technology systems such as aircraft, missiles and space systems, and electronics and communications equipment.[4] As a result, the military's share of industry output in sectors such as aerospace, electronics, and communications is considerably higher than for durable manufactured goods as a whole. Although the military's share of industry output was declining during the 1970s, the Defense Department's purchases have significantly influenced the direction of the high technology industries. Moreover, since these statistics exclude production of nuclear weapons and that portion of the space program with direct military applications, these figures could understate the military's claim on total technological resources by as much as a quarter.

The Pentagon further influenced technological development by funding 38.1 percent of all public and private research and development (R&D) between 1960 and 1973. In the post–Vietnam period, this figure fell to 25.6 percent. Space-related R&D, at least 20 percent of which had direct military applications, averaged another 11.8 percent of all R&D between 1960 and 1973. Since then, the space program has accounted for 7.2 percent of all public and private R&D.

Military R&D has been the federal government's largest mechanism for influencing technological growth. Defense Department R&D averaged 61.4 percent of all federal R&D between 1960 and 1973. Since then it has accounted for 52.7 percent. Space research accounted for another 16.7 percent before 1973 and 14.3 percent since then.

Between weapons procurement and R&D, the Pentagon employs a substantial share of our nation's technical personnel. Estimates of the percentage of scientists and engineers engaged in Defense Department-sponsored projects range from 15 to 50 percent.[5] While the higher estimates might have applied to research and development during the 1950s and 1960s, they could not have covered all production personnel as well.[6]

The most accurate data come from a National Science Foundation (NSF) survey conducted in 1978.[7] NSF's data show that in 1978, 16.2 percent of the nation's engineers and scientists (excluding psychologists and social scientists) worked on national defense as their most important task. Another 3.8 percent primarily worked on space research. The percentages are significantly higher for fields directly related to aerospace and electronics. For example, 60.2 percent of the aeronautical engineers and over 35 percent of electronics engineers worked primarily on national defense. Since NSF's data were gathered for 1978, a year in which the share of GNP devoted to the military was at its lowest point since 1950, there is good reason to believe that the average percentage of all scientists and engineers involved in Defense Department programs over the past two decades is significantly higher.

The available data and estimates by other experts lead us to conclude that 25 to 35 percent of America's scientists and engineers worked primarily on Pentagon projects during the 1960s. During the 1970s, this figure probably dropped to between 15 and 25 percent. As a result of the current military buildup, this percentage is likely to rise significantly during the 1980s.

If the economic benefits of devoting these technological resources to the military outweighed the costs, we would expect to find that the technological superiority enjoyed by American industry during the 1960s would have been maintained or expanded. Since the Pentagon has "seeded" our research laboratories and purchased new products when they were too costly for civilian applications, American industry should have been in an excellent position to commercialize high technology goods. We would also expect that technological advancements resulting from our military effort would have enhanced the efficiency of our factories, leading to increases in manufacturing productivity. Indeed, since America's support for military technology was only part of the largest R&D effort undertaken by a major industrial nation (except the Soviet Union) over the past two decades, there is every reason to expect these results. Unfortunately, neither spinoff has occurred. Since 1960, American high technology industries have lost ground to the Japanese and the Western Europeans in the competition for shares of both the U.S. domestic and worldwide markets. Growth in the productivity of American manufacturers has also fallen substantially.

American firms have experienced some of their largest market-share reductions in industries that are heavily engaged in military contracting, including aircraft, electronics, and machine tools. Although these American high technology industries are growing, they are not keeping pace with competition from abroad in civilian markets. For example, Japanese firms have virtually taken over the commercial electronics market, including televisions, stereos, portable radios and cassette players, and newly introduced video cassette recorders. In 1964, the Japanese did not export color television sets. By 1977, the Japanese had captured 42 percent of the world market and 37 percent of the American market for this product.[8] Japanese control over the video cassette recorder (VCR) market is even more one-sided. While the first video tape machines were produced in the United States for the television industry, the Japanese were the first to develop a marketable consumer version. They currently control virtually all sales of VCRs.[9]

More importantly, the Japanese are making a concerted effort to challenge American preeminence in semiconductors. Development and production of these small silicon chips, which represent the "state of the art" in electronics technology, was dominated by American manufacturers as recently as 1974. Yet since that time, Japanese firms have entered the competition to mass-produce an important segment of this market, memory chips. In this quickly changing field, the Japanese captured 40 percent of the market for the last generation of memory chips—16K RAMs, random access memories that can store over 16,000 bits of information. In 1978, the Japanese introduced the first reliable 64K RAM, representing a fourfold increase in the storage capacity over the 16K RAM. The Japanese currently control about 70 percent of the world market and 50 percent of the U.S. market in 64K RAMs and show every sign of becoming the industry leaders in this technology by being the first to introduce the next generation of memory chips: the 256K RAM.[10]

Similar deterioration of America's technological lead has occurred in machine tools and the emerging field of robotics. As recently as 1967, the United States accounted for 34 percent of the world production of machine tools. However, over the past decade, American firms have not kept pace with the growth of machine tool production in Western Europe. By 1981, U.S. manufacturers were responsible for only 19.5 percent of world output. Western European firms, led by the West Germans, now account for 34 percent of the total.[11]

American manufacturers are also failing to keep pace in the development of emerging machine tool technologies. While computer-controlled machine tools and robotics were American innovations, helped along by Pentagon-sponsored programs, the Japanese now threaten to dominate the commercial application of these technologies during the next decade. Last year, Japanese machine tool builders accounted for over 50 percent of the sales of numerically controlled machining centers in the United States.[12] This is a staggering invasion, considering the fact that they accounted for only 4 percent of the U.S. market in 1974. The situation in robotics is hardly more encouraging. Today, there are 4,500 computer-programmable robots working in the United States, up from 200 in 1970. In Japanese factories, meanwhile, 14,000 robots have already been installed, accounting for 70 percent of the world total.[13]

In the commercial airframe market the European consortium, Airbus Industrie, is challenging Boeing for control of the world market for the new generation of fuel-efficient jumbo-jets. Lockheed has already dropped out of that commercial market, and McDonnell Douglas's DC-10 program continues to suffer substantial losses. Potentially more damaging to America's aircraft industry is its failure to develop an entry to compete in the booming commuter airline market. The Congressional Office of Technology Assessment estimates that by the year 2000, commuter airlines will order 6,000 new aircraft. Yet, in a pattern reminiscent of the U.S. auto industry in the 1960s, American manufacturers, including Beech, Cessna, and Piper, have failed to invest in the technology necessary to develop an aircraft that can compete effectively in that market. Consequently, America's commuter airlines are turning to Canadian, French, and Brazilian firms to fill their needs.[14]

America's declining productivity growth also suggests that our technological progress has slowed. Two of the more important paths to greater manufacturing efficiency are improving production technology and creating more attractive products. Funding for research and development should thus have a positive impact on productivity growth because new tools and new products are nurtured in our research labs. While traditionally the relationship between R&D and productivity growth has been found by economists to be quite significant, recent evidence suggests that the relationship has changed.[15]

Cross-national data comparing R&D expenditures are indicative of this change. While American investment in R&D has been substantial, productivity growth has been weak. The United States has maintained the greatest number of R&D scientists and engineers and the highest proportion of these researchers in the total labor force of any country except the Soviet Union. We spend more on R&D than France, West Germany, and Japan combined.[16] In addition, U.S. expenditures on R&D as a proportion of GNP were higher than four of the five other industrial countries for which data can be obtained—the United Kingdom, France, West Germany, and Japan. The Soviet Union was again the exception. Yet in spite of this enormous R&D effort, U.S. productivity growth during the 1970s ranked second to last, among those six nations.

If we chart the share of national resources that the six nations devoted to military and space R&D, we find that as that factor increases productivity growth tends to decline. The Soviet Union, the United States, and the United Kingdom are at the top of the list in military-related R&D expenditures and at the bottom in productivity growth. It seems that while those nations have been locked in a technological arms race, Japan, West Germany, and France have been concentrating on developing civilian technology that increases manufacturing efficiency.

One possible reason that military R&D does not seem to stimulate, and may even hinder, productivity growth is that only a small fraction of America's military R&D is spent on basic research. Technological exploration undertaken to expand the frontiers of science is an important source of innovation. The Defense Department, however, has spent only about 3 percent of its R&D funds for such purposes during the past two decades.[17] While the Pentagon spent almost half of all federal R&D dollars in 1980, three other federal agencies spent more funds on basic research—the Department of Health and Human Services, the National Science Foundation, and the National Aeronautics and Space Administration.

Although data on the loss of market shares and productivity growth are only suggestive, they cast doubt on the proposition that military spending helps technology more than it hinders it. We recognize that numerous factors influence international competitiveness and production efficiency. Yet the trends of those data support the thesis that the negative effects of military spending on technology outweigh the positive spinoffs. "The decline in productivity and industrial standards in the U.S.," commented one Japanese observer, "is the best argument against the idea that more defense contracts are vital to maintaining 'state of the art' efficiency."[18]

The reason that military spending has probably slowed our technological progress seems clear: using scientific and engineering talent to solve military problems is an inefficient means of stimulating scientific or commercial advancement. Growth in our base of scientific knowledge comes most readily from basic research without the constraint of specific applications. The development of new products, like fuel-efficient automobiles, alternative energy systems, and computer-controlled machine tools, is most quickly accomplished by applying R&D talent directly. While military programs sometimes provide a market for new products and occasionally result in a civilian spinoff, much of the effort expended to develop weapons systems, like laser-guided missiles and electronic jamming devices, does not help the civilian economy. As one of the founders of TRW Corporation, Simon Ramo, puts it:

> [T]he fallout [from military spending] has not been so great as to suggest that for every dollar of military technology expenditure we realize almost as much advance of the non-military fields as if we had spent it directly on civilian technology. Probably our relative productivity increases and our net rating in technology vis-à-vis other nations have on the whole been hurt rather than helped by our heavier involvement in military technology as compared with other nations.[19]

The nature of military spending and the Pentagon's spending patterns heavily contribute to the inefficiency. Since the military stresses high performance over cost, technologies developed for the military are often too expensive and complex for widespread civilian use. Many of those that do have civilian application, like radar and nuclear power, had to undergo significant redesign before they were commercially viable. Military contracting also tends to favor larger firms that are traditionally less innovative and create fewer new jobs than smaller enterprises.[20] Since the 1950s, about 80 percent of all military contracts (including R&D and procurement) have gone to large firms.[21]

Moreover, military requirements can distort engineering practices by placing greater emphasis on high-performance capabilities than on reducing cost. Evidence of this problem can be found in the difficulties that military contractors have experienced attempting to develop civilian products. For example, attempts to enter the mass transit market by Boeing Vertol (trolley cars), Rohr (subway cars), and Grumman (buses) all failed in part because their products were too complex and unreliable.

We shall now explore the military's impact on technology in greater detail through an examination of the Pentagon's role in the electronics industry.

CASE STUDY: ELECTRONICS

The revolutionary advances in electronics over the past few decades have resulted in products that our great-grandparents would never have dreamed of. These new devices, including computers, hand-held calculators, word processors, and video cassette recorders, have significantly changed our lives. At work and during our leisure time, these innovations have improved our access to information and entertainment; expanded our ability to analyze, process, and store data; and increased our ability to control machines and energy use. In connection with new machine tool technologies such as robotics and computer-aided design and manufacturing (CAD/CAM), semiconductor-based electronics might very well transform the factory within the next decade as thoroughly as mechanization and transportation changed agriculture during the last hundred years. As one employee of a major semiconductor firm sees it, we have only begun to recognize the possible applications for this new technology: "Any product that uses springs, levers, stepping motors or gears is performing logic and that product should be built of semiconductors."[22]

There is no doubt that the military has helped stimulate the electronics industry's technological growth by acting as a "creative first user" of the industry's products. The Pentagon has purchased innovations before they were commercially marketable and subsidized efforts to improve their quality and reliability. For example, government sales were by far the most important market for integrated circuits when they were introduced in the early 1960s, and the percentage of electronic sales in the United States accounted for by the government, which were overwhelmingly military and space-related, accounted for over 40 percent of the total between 1952 and 1969.

In addition to subsidizing high technology products, it has been argued, the development of military hardware has also resulted in concepts, basic knowledge, and skills relevant to the civilian arena. As two Massachusetts Institute of Technology scientists explain:

> [M]ilitary requirements initiated other developments. Missile guidance systems were an early source of support for integrated circuit development; requirements for satellite-tracking radars have supported the development of surface acoustic-wave technology and charge-coupled devices, as well as modern signal processing techniques. . . . The need for reliable military communications has led to the . . . development of satellite communications [and] to small, mobile terminals. On-line computer time-sharing, computer networks, and computer graphics were all initially supported by the military. Finally, we should mention such significant second-order developments as radio and radar astronomy, microwave spectroscopy, and instrumentation for earth-resources satellites and for modern health care, all of which are heavily dependent on concepts and components derived from military electronics.[23]

This broad claim raises serious questions: How should we evaluate the technological impact of military spending? Should we be thankful for *any* civilian advance resulting from military research and production? Or should we weigh the costs against the benefits of these expenditures? However these questions may be answered, we have concluded that *unnecessary military expenditures can never be justified on the basis that they will create an indeterminate amount of technical knowledge transferable to the civilian sector.* While we recognize that military programs significantly contributed to the early development of the electronics industry, our assessment indicates that the costs now seem to outweigh the benefits. Four factors are particularly important:

1. The end-products developed for military use, such as missile guidance systems and electronic warfare devices, have few commercial applications.
2. Military research and development has a mixed record in supporting projects that have lead to significant innovations. The difficulty seems to stem from two factors. One, it is hard for military personnel to judge the most promising projects. Two, the Pentagon tends to support larger, established firms, rather than the younger, smaller firms that have proved more innovative.
3. Military subsidies for new products are increasingly unnecessary. As the cost of electronic circuits has rapidly fallen, massive industrial and commercial markets have overshadowed the military's role as "creative first user."
4. The Pentagon's current attempts to guide the development of semiconductor and machine tool technology raise serious economic and political issues. Military subsidies for new semiconductor technology in the 1980s may hinder development in this field rather than nurture new products. Defense Department programs designed to increase factory automation are financing a narrow approach without input from the workers who will be affected.

Few Transfers

There are only a few examples of electronic devices originally designed for the military that have later found civilian application. Of those transfers, such as navigational radar, substantial redesign was required for the product to meet the far less complex civilian requirements.[24] This barrier arises from the basic differences between the military and commercial environments: there is little sensitivity to price within the military arena, while cost is the central concern of the commercial customer.[25] In addition, the custom-made requirements of the Pentagon conflict with the civilian demand for standardization.[26] Thus, as military technology becomes more costly and complex, fewer civilian applications can be found. As the *Wall Street Journal* observed of the aerospace industry as early as 1963:

> When the government was creating the old piston-powered warplanes . . . this technology could be easily and profitably translated into commercial airliner designs. When the military moved on to jet planes, the translation became far more difficult and costly—as the financial plight of both manufacturers and airlines currently testifies. The supersonic military jet has thus far defied any commercial translation at all, and the government now debates whether it is worth pouring tax money into subsidizing creation of a supersonic airliner. As for commercial transport application of a moon rocket, it defies even day-dreaming.[27]

The trend is similar in the electronics industry. Night-vision devices, radar jamming equipment, and missile guidance systems are all examples of technologies that promise only limited civilian applications. Even early integrated circuits developed for the Minuteman II missile "attracted little attention because of their distinct military application."[28]

The "cultural" differences between the military and civilian markets were identified by a British researcher as key reasons for the limited diffusion of military innovations in that country. Sir Ieuan Maddock, in a report to the U.K.'s Electronics Economic Development Committee, found that defense firms lack the entrepreneurial skills required for civilian production. British defense firms, his interviews revealed, view nonmilitary products as "unworthy of the high quality scientists and engineers they employ"; worry about competing with stronger civil technology firms; are unfamiliar with the need for shorter development periods; are frustrated by the lack of clearly defined product requirements; and are troubled by the fact that no "long term guarantee" for the purchase of a product can be found in the civilian arena.[29]

These "cultural" differences are much the same in the United States. Yet largely because of greater labor mobility and a bigger domestic market in this country, technologies and components first developed by the military have found their way into civilian products.[30] On the whole, however, products developed for military applications have not been transferable to the civilian sector and thus have acted as a drag on the nation's technical resources.

Inefficient R&D

Military research and development in semiconductors has generally not produced basic technological innovations. Although the military dominated R&D spending during the last three decades, numerous studies indicate that far fewer patents resulted from these efforts than from commercially funded projects. Moreover, of the patents that did result from military R&D, very few found commercial application.[31] In the words of one analyst: "Texas Instruments reported that between 1949 and 1959, only five of 112 patents awarded to the company were developed under government contract, although the government funded two-fifths of R&D spending. Further, only two of the five patents were used commercially."[32]

Counting patents is somewhat misleading because the military usually retains some control over innovations developed with Pentagon support. Thus, firms have an incentive to finance potentially profitable innovations privately.[33] Yet it still stands to reason that R&D sponsored by a firm attempting to develop a specific product aimed at the commercial market would be more productive than research designed to solve a military-related problem that may or may not have other applications. As Robert Noyce, a key figure in the development of semiconductor technology, observed:

> With very few exceptions, the major motivation behind technology development cannot come from the military . . . the major motivation, I feel, is the commercial one. . . . I would say that the military created more motivation for doing good research by creating a market for advanced products. . . .[34]

Noyce's observations hold particularly true for the development of semiconductors. None of the major innovations in this field was directly supported by military research; however, many of the new technologies were heavily subsidized by the military through substantial early purchases. For example, the scientists who first discovered the transistor at American Telephone and Telegraph's Bell Labs in December of 1947 were driven, in large measure, by the pure scientific challenge of understanding materials that sometimes did, and sometimes did not, conduct electricity. More pragmatically, John Bardeen, Walter Brattain, and William Shockley were searching for a simple electronic switch to replace bulky mechanical devices that limited the size of the telephone network.[35] The inventors were so concerned that the military would classify information about their innovation that they did not reveal the transistor's secret to the military until they were certain that it would become public knowledge, allowing Bell to develop its civilian applications.[36]

Ironically, the military, needing smaller, quicker, and less power-consuming electrical components, soon became the largest purchaser of transistor devices. Another industry that provided a market for the early, very expensive transistor was hearing-aid manufacturers. These firms shared with the military a need for a more compact amplifier than the vacuum tube. The transistor was the perfect answer. Customers for hearing-aids were also generally willing to pay a premium price.[37] Nevertheless, during the early 1950s, when the transistor was unreliable and expensive to manufacture, the military consumed the vast majority of the infant industry's output.

By the late 1950s, when the two innovations occurred that made it possible to place numerous transistors on one silicon chip, the military was heavily engaged in research toward this goal. As it turned out, however, *none* of the three different projects supported by the three military services was instrumental in these discoveries. Neither Texas Instruments' (TI) discovery of the integrated circuit, linking many electrical components on one device, nor Fairchild Semiconductor's development of the planar process for mass-producing silicon chips, was supported by the military.[38] Yet, as in the case of the transistor, the government market for integrated circuits provided the production experience necessary to make these devices commercially attractive. The Minuteman II project subsidized TI's early integrated circuit production and the NASA Apollo program played the same role for Fairchild.[39]

The military's failure to "back the right horse" in the quest to miniaturize electronic components suggests that military-sponsored programs are a less efficient means of stimulating technological innovation than private efforts. Pentagon-sponsored research aimed at miniaturizing electronic components was spurred on by the Soviet Union's successful Sputnik satellite launch.[40] The Army Signal Corps spent $26 million between 1957 and 1963 on the "micromodule" program designed to stack and encapsulate transistors. This program was seen as an attempt to spur the evolution of the transistor. Much of the research was carried out by RCA, a large, old-line electronics firm. The program, which was only marginally successful, ended as the integrated circuit and the planar process became the accepted miniaturization technique.[41]

The air force took another approach. They pursued a revolutionary breakthrough in "molecular electronics." Jack Kilby, inventor of the integrated circuit at TI, described the air force's project: "The Air Force rejected anything that had any connection with existing circuits. . . . They didn't want to get there circuit by circuit. They wanted these new breakthrough devices that would eliminate all that jazz."[42] Westinghouse, another large electronics firm, received the molecular electronics contract. One industry analyst has suggested that Westinghouse won the contract because "no other company was interested in taking such a leap into the technological dark."[43] While some have argued that the "micromodule" program increased the industry's interest in miniaturization, the research carried out by Westinghouse under this program did not result in any technological breakthroughs.[44]

The navy backed "thin film" research designed to "print" a circuit and passive components, like resistors, onto a thin ceramic wafer. While this work did not directly contribute to the discovery of integrated circuits, it did advance photoresist techniques. Yet, by and large, military-funded research and development was not particularly helpful in the search for methods to miniaturize electronic circuits.[45]

One explanation for this failure, suggested by industry executives, is that the military officers in charge of these programs did not fully understand the technology.[46] This would not be surprising, considering the speed of change in the semiconductor industry. Yet this is probably not the only reason because venture capitalists and industry executives also had difficulty forecasting the direction of semiconductor technology.

Another important factor that may have neutralized the effectiveness of military R&D was the spending pattern: most of the money went to projects carried out by larger electronics manufacturers. In 1959, for example, 78 percent of government R&D funds went to the eight established vacuum tube producers (by then, also producing transistors) and Western Electric (ATT's manufacturing division). Yet those firms accounted for only about 37 percent of the semiconductor market. New companies entering the field, which by 1959 had captured 63 percent of the market, received only 22 percent of the government's R&D expenditures.[47]

The government's funding pattern ran counter to the trends that made the American semiconductor industry so successful. Most observers agree that the mobility of technical information and personnel, and the ease of starting a new firm were key ingredients in the rapid technological growth of the U.S. semiconductor industry.[48] Established firms were slow to catch on to the potential of the transistor. At the same time, Bell Labs, concerned about antitrust action, provided relatively easy access to technical information and did not prevent professional staff from leaving to start new commercial ventures. One such person was Gordon Teal, who demonstrated the first silicon transistor after leaving Bell to head the research arm of TI. In describing the failure of European firms to capture a significant share of the semiconductor market, two British observers suggested that "much of the pace of semiconductor development in the United States can be explained in terms of the decline of the old and the rise of the new. . . ."[49]

Indeed, as the technology progressed into the 1970s, this continued. Metal oxide semiconductors (MOS), which made possible low-cost calculators, watches, and large-scale semiconductor memories (RAMs), were pioneered by new firms like Mostek (established in 1969) and Intel (1968).[50] Firms such as TI and Fairchild, now part of the establishment, were convinced that the technology would move in another direction. They did not vigorously pursue this technology until after it was demonstrated by the "upstarts."

The more established firms in the industry were convinced that the integrated circuit industry would evolve toward providing "custom" designs for the final product manufacturers, like computer firms.[51] But the development of MOS technology, and Intel's 1972 introduction of the microprocessor, propelled the industry in two separate directions that vastly reduced the importance of the custom market. As mentioned before, MOS made possible numerous highly profitable standard products, such as the calculator chip first developed by Mostek in 1972. The microprocessor, on the other hand, "was a pioneering advance in *flexible* product design which gave the industry a new way out of the custom versus standard battle: By programming the on-chip memory, the microprocessor could be customized for each application."[52]

Neither MOS nor microprocessor research was supported by the Pentagon. A calculator firm supported both Mostek's development of the calculator chip and Intel's microprocessor work. This support foretold a major change in the semiconductor industry: civilian markets were destined to overshadow the military. Indeed, support from the military, in the form of early purchases of expensive models and assistance aimed at improving the production process, proved significantly less important to this new generation of products than it had been to the integrated circuit and the planar process.[53]

Decreasing Importance of Military as First User

During the late 1970s, numerous analysts concluded that military spending had become much less important to technological progress in electronics than it had been during the 1950s and 1960s.[54] Two observers argued that the Pentagon's role diminished because the increasing emphasis on cost at the Pentagon lessened the support for expensive, emerging innovations.[55] This assertion, however, falls short of explaining the military's changing role. Current cost overruns on major weapons systems are as bad as they have been throughout the cold war. Moreover, the Pentagon's most profitable support for semiconductor production, the Minuteman II program, was carried out under a Secretary of Defense, Robert McNamara, who instituted numerous reforms designed to reduce costs.[56]

Other explanations for declining impact of military spending include: (1) increasing reliance on established suppliers, at the cost of ignoring smaller innovative firms; (2) the decreased willingness on the part of all government agencies to consider unsolicited proposals; and (3) the expanding paperwork burden imposed by Pentagon contracting procedures.[57] The first argument clearly does not stand up. Defense Department support for small businesses has remained remarkably stable but low, at about 20 percent of total contract awards since the 1950s. In fact, the share of contract awards going to small businesses is actually greater today than it was during the late 1950s and early 1960s.[58]

Although the other two factors may have had some impact, each of these explanations overlook the fact that semiconductors have become so cheap and standardized that they can be profitably used in a wide array of civilian products. In the last decade, civilian demand for semiconductors has become large enough that this market, rather than the military, drives technology advancement. Moreover, as the semiconductor industry entered the 1970s, the military's importance as a market declined. The government's share of semiconductor output fell from 36 percent of the total in 1969 to about 10 percent in 1978.[59] This reduction occurred, not so much as a result of falling government demand, but rather because of the explosive growth in the use of semiconductors in industrial and consumer products during that period. As the cost of semiconductor products fell, numerous civilian applications were found. Microprocessors were adopted to control a broad variety of machines, from heating and cooling equipment in large office buildings and factories to telecommunications systems and machine tools. Memory chips were quickly adopted for use in word processing machines, home computers, and video games.

With the dramatic reduction in the price of semiconductor products and the fast-paced growth of civilian markets, military subsidies for new semiconductor innovations have become much less important. As we mentioned before, the introduction of microprocessors and MOS memories took place without significant government purchases. Microminiaturization, known as Very Large-Scale Integration (VLSI), has also been pushed along primarily by competition for civilian markets. Indeed, the military is so concerned that they may not be keeping up with the growth of semiconductor technology, that they have initiated a program to encourage producers to develop chips specifically designed for military requirements.[60] As a report to the Joint Economic Committee of the U.S. Congress pointed out, ". . . the military's needs are not in the mainstream of the industry's evolution."[61]

Japan's recent success in one of the most sophisticated parts of the semiconductor market is clear evidence of this assessment. Until just recently, American semiconductor producers were unchallenged in the world market. Although the Japanese produced transistors and later, semiconductors for export in consumer electronics products, they exported virtually no semiconductors for computers and other industrial purposes until 1974.[62] Yet when they did begin competing for a part of the international market for a sophisticated product—memory chips—the Japanese soon became a dominant force.

In 1978, when the introduction of IBM's series 4300 computer helped create a sudden surge in demand for memory chips that could not be met by American producers, Japanese companies were able to capitalize quickly on the shortfall and capture 40 percent of the market for advanced memories—16K RAMs.[63] Japanese firms also introduced the first commercially successful 64K RAM device and established an early market lead. Further, they are planning to introduce the next generation of memories—the 256K RAM—within a year. While some observers feel it is too early to tell how large the Japanese share of the U.S. memory market will be, at least one leading semiconductor executive believes that the battle is over and the Japanese have won.[64]

The Japanese have proved that commercial development of an advanced semiconductor product is possible without military subsidies. Still, the growth of Japan's semiconductor industry did not occur without government support. Since the 1960s, Japanese government agencies, particularly the Ministry of International Trade and Industry (MITI), have worked cooperatively with Japanese firms to make the development of knowledge-intensive industries a national goal. A recent MITI document is explicit about Japan's aim:

> It is extremely important for Japan to make the most of her brain resources, which may well be called the nation's only resource, and thereby to develop creative technologies of its own. . . . Possession of her own technology will help Japan *to maintain and develop her industries' international superiority* and to form a foundation for the long-term development of the economy and society. . . .[65] (emphasis in original)

Three elements were particularly important in the Japanese strategy for expanding the semiconductor industry:

1. *Protection of the domestic electronics industry from American products until the mid-1970s.* In 1968, American firms controlled less than 15 percent of the Japanese semiconductor market.[66] The Japanese also prevented American companies from setting up production facilities in Japan, with the exceptions of TI and IBM (TI's Japanese subsidiary produces semiconductors and IBM's builds computers).[67] Limited access to the Japanese market allowed domestic firms to develop their expertise in the advanced semiconductor industry by providing a built-in market for Japanese products. While consumer product exports subsidized the semiconductor industry's early development, the Japanese needed protected computer and telecommunications markets to move toward the technological frontier.[68]

2. *Purchase of U.S. technology and production equipment.* To gain experience in this field, Japanese firms purchased American semiconductor technology and production equipment. American

firms settled for royalty payments on their patents after realizing that they could not penetrate the Japanese market. By the end of the 1960s, at least 10 percent of the Japanese semiconductor industry's revenues were paid as royalties to U.S. firms.[69]

3. *Support for, and coordination of, research and development efforts.* Since 1971, Japan's R&D efforts have been aimed at developing semiconductors for the computer market. MITI encouraged three paired groups of Japan's six dominant semiconductor firms to pursue different aspects of Large-Scale Integration (LSI) technology in hopes of capturing a share of the international market for the next generation of computers. Each of the three groups received about $200 million in subsidies between 1972 and 1976.[70] Through this program, the Japanese were able to displace American-developed semiconductor products from their domestic market in all but the most advanced applications. They also succeeded in significantly raising the domestic share of installed computers.[71]

In December 1975, when the Japanese had significantly closed the technology gap, the Japanese government began to liberalize trade policy. While there is skepticism about how far formal liberalization actually opened the Japanese market,[72] it did set the stage for full-scale Japanese participation in the international semiconductor arena. In 1976, the Japanese semiconductor industry and government created the VLSI Technology Research Association to insure that they could successfully compete in that market. This grouping, which included MITI, the Japanese Telephone Company (Nippon Telephone and Telegraph) and five major semiconductor firms, coordinated research on VLSI technology and shared results to prevent duplication. Firms within the group received significant subsidies that, among other benefits, released company funds to invest in production capacity. Through these mechanisms, the Japanese semiconductor industry was quickly able to increase its technological capacity and successfully compete with American firms.

The success of Japan's basic strategy suggests that in the future, military spending will be an ineffective mechanism for competing with Japanese firms in the high technology arena. While other factors, such as the availability of capital and lower labor costs, have also contributed to Japan's success, the cooperation between industry and government to develop a technology with wide commercial application has proved quite effective. Particularly in semiconductor technology, which has matured sufficiently so that there is widespread recognition of its applications as well as significant production experience and a degree of standardization, there are likely to be few instances where the military's most effective role—creative first user—will be helpful in supporting technological innovation. Indeed, in the semiconductor field the possibility of the military hindering progress, by diverting resources from the civilian realm, is greater than the likelihood of it stimulating growth.

While our antitrust laws and trade policies prevent many of the techniques used by the Japanese from being copied in the United States, we still should be able to find more effective ways of stimulating technological growth in that field than relying on military R&D. In the United States, it might make sense to expand government support for basic research at universities in fields applicable to semiconductor technology. Loan guarantees for companies developing promising innovations might be another method for achieving this goal. While more specific recommendations, or a more detailed assessment of the reasons for Japan's success, are beyond the scope of this study, our overview indicates that the military is no longer the most effective way for the government to encourage growth in the semiconductor industry.

Distorting Technology

After watching its influence slip away during the 1970s, the military will attempt to expand its involvement in the high technology sector during the next few years. As in the past, the vast majority of the military's programs are oriented toward developing existing technologies rather than adding to basic knowledge. The programs that are scheduled to grow raise far-reaching economic and social questions. Two efforts that deserve close scrutiny aim at: (1) developing quicker integrated circuits for "smart" weapons, and (2) creating increasingly automated production equipment to reduce the cost of weapons systems. Further Defense Department involvement in the semiconductor field could actually hamper attempts to stay even with the Japanese by diverting technically skilled workers from civilian projects. Military programs designed to automate the factory are financing a narrow approach to automation without significant input from those who will be affected.

As the American electronics industry has grown, the military's role as primary customer has declined. So too has its power to direct technological development. To counteract this declining influence, the Defense Department launched the Very High Speed Integrated Circuit (VHSIC) program. The Under Secretary of Defense for Research and Engineering explained the need during budget hearings in 1980:

> Between our funded Science and Technology program and the contractors' independent R&D program, we have an outstanding ability to direct technology resident in the defense industry to high priority defense programs. However, we have little ability to influence those companies whose sales are predominantly commercial. This is a serious limitation in the case of the semiconductor industry, whose products play a crucial role in nearly all of our advanced weapon systems. Therefore, we have initiated a new technology program intended to direct the next generation of large-scale integrated circuits to those characteristics most significant to defense applications.[73]

The VHSIC program is designed to generate interest in the semiconductor industry for the development of chips with 10 times the density and 100 times the speeed of current chips. The military needs these faster and smaller chips for such sophisticated weapons as precision-guided munitions, air-to-air missiles, cruise missiles, ICBMs, night-vision devices, and torpedoes. While further research on miniaturization could help the commercial sector, many firms have expressed concern that the VHSIC program will divert resources from civilian efforts and retard technological progress.[74] As one industry executive put it: "Those incredibly complex VHSI circuits the military wants don't seem to have their use elsewhere."[75]

While the military's involvement in the semiconductor field could reduce the competitiveness of American firms, the Pentagon's support for research on factory automation raises a different kind of concern. Is the Defense Department the correct agency to control public funds destined for a program that could have widespread impact on work in America?

In 1973, the Pentagon initiated an effort to develop a fully automated "factory of the future." This program, known as Integrated Computer-Aided Manufacturing (ICAM), is exploring ways to coordinate the work of a factorywide network of machine tools and robots through computers. If successful, ICAM could significantly reduce the number of skilled machinists needed in defense facilities.

The jobs that do remain may require far fewer skills. For example, one ICAM prototype facility was able to process 25 percent more material with 44 percent fewer workers.[76]

While automation can increase productivity, leading to better economic performance, there are many ways that mechanization can proceed. Mechanization can call for part of the existing workforce to enhance its skills, or it can leave those remaining workers with menial tasks. In either case, automation is becoming an increasingly critical issue to both management and labor. In the ICAM program, the Defense Department has backed management's vision of the future factory without equivalent consideration of labor's position. Labor has had virtually no say about the philosophy or design of the Pentagon's manufacturing technology programs. Moreover, while the ICAM prospectus lists numerous program goals—including better productivity, increased profits, and greater design flexibility—the goal that industry participants consider to have the highest payoff potential is greater management control over the production process.[77]

Right now, machine tool operators are very skilled and with that skill comes a great deal of autonomy. While machinists do not choose what is to be produced, they can exert control over numerous factors, including the pace of work. The fact that their work is complex makes it more interesting. It also means that they must be paid more than less skilled workers.

There is nothing surprising about an industry seeking to increase management control over production. But Pentagon subsidies for such programs raise at least two major issues. First, should the government assist one segment of the society without hearing from those who will be affected? Our view is clear. All voices should be considered in decisions made by a representative government. The ICAM program goes forward primarily because the shroud of "national security" sometimes allows programs to proceed that would not otherwise pass the normal tests for a sound public policy. Second, is the Defense Department the best agency for nurturing new manufacturing technology? As we have seen, the Defense Department has sometimes assisted the development of new technology. However, the military's orientation toward performance rather than cost could push ICAM efforts into areas just as irrelevant to manufacturing technology as the army's "micromodule" research was to semiconductor development.

Both ICAM and VHSIC illustrate the problems created by allowing the Department of Defense to control most of the government's impact on industry and technology. While both programs are designed to assist industry in developing new technology, the final objective of these developments is to create new weapons. In sharp contrast, technological progress subsidized by the Japanese government is carefully targeted to commercial opportunities. As a result, the Japanese have dramatically expanded their technological expertise within a very short time. In this country, the least questioned reason for the government to become involved in private business is to produce weapons. Yet this may prove disastrous if we want to remain a world leader in technology. Maybe we should consider developing a civilian-oriented technology program and hope some of the results are useful to the military.

NOTES TO CHAPTER 5

1. Simon Ramo, *America's Technology Slip* (New York: John Wiley & Sons, Inc., 1980), pp. 79–80. G. P. Dineen and F. C. Frick, "Electronics & National Defense: A Case Study," in P. Abelson and A. Hammond, eds., *Electronics: The Continuing Revolution* (Washington, D.C.: American Association for the Advancement of Science, 1977), pp. 82–83.
2. David Blond, "On the Adequacy and Inherent Strengths of the United States Industrial and Technological Base: Guns versus Butter in Today's Economy" (Office of the U.S. Secretary of Defense, Program Analysis and Evaluation, 1981). (Mimeo.)
3. James M. Utterback and Albert E. Murray, "The Influence of Defense Procurement and Sponsorship of Research and Development on the Development of the Civilian Electronics Industry" (Center for Policy Alternatives, Massachusetts Institute of Technology, Cambridge, Mass., June 30, 1977), p. 3.
4. U.S. Department of Defense, Washington Headquarters Service, *Prime Contract Awards, FY1981* (February 1982), Table 6.
5. Jacques S. Gansler, *The Defense Industry* (Cambridge, Mass.: MIT Press, 1980), p. 54, estimates that 20 to 30 percent of all scientists and engineers work on military projects. Richard Dempsey and Douglas Schmude, "Occupational Impact of Defense Expenditures," *Monthly Labor Review* (December 1971): 12, estimate 20 percent. Murray L. Weidenbaum, *Economics of Peacetime Defense* (New York: Praeger, 1974), p. 27, estimates 50 percent. Ramo, *America's Technology Slip*, p. 80, estimates 50 percent.

6. "Fifty-two percent of the engineers and scientists engaged in research and development work in American industry in 1963 were working on projects financed by defense or space programs," from William L. Baldwin, *The Structure of the Defense Market* (Durham, N.C.: Duke University Press, 1967), p. 146.
7. National Science Foundation, *Characteristics of Experienced Scientists and Engineers 1978* (Washington, D.C.: N.S.F., 1979).
8. Ira C. Magaziner and Robert B. Reich, *Minding America's Business: The Decline and Rise of the American Economy* (New York: Harcourt Brace Jananovich, 1981), p. 169.
9. Electronic Industries Association, author's phone conversation with a public affairs officer, October 1982.
10. Andres Pollack, "Japan's Big Lead in Memory Chips," *The New York Times*, February 28, 1982, p. F1.
11. *The American Machinist* (February 1982): 107.
12. National Broadcasting Corp., Inc., Transcript, *NBC Reports: Japan vs. USA, The High Tech Shootout* (New York: NBC, August 14, 1982), p. 34.
13. Robert Ayres and Steve Miller, "Industrial Robots on the Line," *Technology Review* (May/June 1982): 38.
14. "Study Is Critical of U.S. Aircraft Makers," *The New York Times*, February 22, 1982.
15. Eleanor Thomas, "Recent Research on R&D and Productivity Growth: A Changing Relationship Between Input & Impact Indicators?" (Washington, D.C.: N.S.F., September 1980).
16. National Science Board, *Science Indicators 1980* (Washington, D.C.: N.S.F., 1981), p. 2.
17. Ibid., p. 270.
18. Tracy Dahlby, "Can the U.S. Really Guarantee Our Security in Wartime?" *Far Eastern Economic Review* (December 5, 1980): 55–56.
19. Ramo, *America's Technology Slip*, p. 251.
20. U.S. Department of Defense, *Prime Contract Awards*, Chart II.
21. Richard S. Morse, "The Role of New Technological Enterprises in the U.S. Economy: A Report of the Commerce Technical Advisory Board to the Secretary of Commerce" (Washington, D.C.: Commerce Department, January 1976).
22. Floyd Kvamme of National Semiconductor, Inc., quoted by Ernest Braun and Stuart MacDonald, *Revolution in Miniature: The History & Impact of Semiconductor Electronics* (London: Cambridge University Press, 1978), p. 120.
23. Dineen and Frick, "Electronics & National Defense," pp. 82–83.
24. Utterback and Murray, "The Influence of Defense Procurement," p. 25.

25. Jacques S. Gansler, *The Defense Industry*, ch. 3.
26. See semiconductor industry executives' negative reaction to military contracts due to the custom nature of the Pentagon's demands in Robert Wilson, Peter Ashton, and Thomas Egan, *Innovation, Competition and Government Policy in the Semiconductor Industry* (Boston: Charles River Associates, March 1980), pp. 6-17.
27. *Wall Street Journal*, May 6, 1963,cited in William Baldwin, *Defense Market*, p. 145.
28. Anthony M. Golding, "The Semiconductor Industry in Britain and the United States: A Case Study in Innovation, Growth, and the Diffusion of Technology" (Ph.D. dissertation, University of Sussex, 1971), excerpted in Norman Asher and Leland Strom, "The Role of the Department of Defense in the Development of Integrated Circuits" (Arlington, Va.: Institute for Defense Analysis, 1977), p. 61.
29. Sir Ieuan Maddock, "Commercial Exploitation of Defence Technology" (London: Electronics Economic Development Committee, 1982), pp. 4-5.
30. Ibid., pp. 7-8.
31. Utterback and Murray, "The Influence of Defense Procurement," p. 24.
32. Mary Kaldor, *The Baroque Arsenal* (New York: Hill and Wang, 1981), p. 91.
33. Semiconductor executives were also reluctant to undertake government-funded R&D programs because of the government's practice of claiming title to any patents that resulted. Wilson, Ashton, and Egan, *Semiconductor Industry*, pp. 6-26.
34. Braun and MacDonald, *Revolution in Miniature*, p. 142.
35. Ibid., pp. 40-42.
36. Ibid., p. 52.
37. Ibid., pp. 55-56 and 69.
38. Ibid., pp. 107-122.
39. Utterback and Murray, "The Influence of Defense Procurement," pp. 11-12.
40. Braun and MacDonald, *Revolution in Miniature*, p. 107.
41. Ibid., pp, 107-110.
42. Ibid., p. 110.
43. Ibid., p. 110.
44. Ibid., pp. 107-110.
45. Ibid., p. 108.
46. Ibid., pp. 109, 136.
47. Ibid., p. 81.
48. Wilson, Ashton and Egan, *Semiconductor Industry*, pp. 2-10, 6-48, 6-65, and 6-66.
49. Braun and MacDonald, *Revolution in Miniature*, p. 154.

50. Michael Borrus, James Millstein, and John Zysman, "International Competition in Advanced Industrial Sectors: Trade and Development in the Semiconductor Industry" (Washington, D.C.: U.S. Congress, Joint Economic Committee, February 1982), Table 3, pp. 30-31.
51. See analysis of the misguided custom LIS venture in Wilson, Ashton, and Egan, *Semiconductor Industry*, pp. 4-16 and 4-21.
52. Borrus, Millstein, and Zysman, "International Competition," p. 28.
53. OECD, "Gaps in Technology: Electronic Components" (Paris: Organisation of Economic Cooperation and Development, 1968), a report excerpted in Norman Asher and Leland Strom, "The Role of the Department of Defense in the Development of Integrated Circuits" (Arlington, Va.: Institute for Defense Analysis, 1977), pp. 36-42. Borrus, Millstein, and Zysman, "International Competition," discuss the reduced role of the Pentagon in the semiconductor market in the 1970s, pp. 32-43, 151-167.
54. Utterback and Murray, "The Influence of Defense Procurement," pp. 47-49. Borrus, Millstein, and Zysman, "International Competition," pp. 10, 151. William L. Baldwin, *The Impact of Department of Defense Procurement on Competition in Commercial Markets: Case Studies of the Electronics & Helicopter Industries* (Washington, D.C.: Federal Trade Commission, Office of Policy Planning, December 1980), pp. 74-77.
55. Utterback and Murray, "The Influence of Defense Procurement," pp. 47-48.
56. Gordon Adams, Paul Murphy, and William Rosenau, *Controlling Weapons Costs: Can the Pentagon Reforms Work?* (New York: Council on Economic Priorities, 1983), pp. 17-18.
57. Morse, "New Technological Enterprises," pp. 4-7. Utterback and Murray, "The Influence of Defense Procurement," pp. 47-48.
58. U.S. Department of Defense, *Prime Contract Awards*, Chart II.
59. Asher and Strom, "Integrated Circuits," Table 4-8.
60. Baldwin, *Case Studies*, pp. 92-93.
61. Borrus, Millstein, and Zysman, "International Competition," p. 151.
62. Ibid., p. 1.
63. Ibid., pp. 105-106.
64. "I think we've already lost out in the 256K," said W. J. Sanders 3d, chairman and president of Advanced Micro Devices of Sunnyvale, California, "The Japanese have won the dynamic RAM market." Quoted in "Japan's Bid Lead in Memory Chips," *The New York Times*, February 28, 1982, p. F1.
65. Ministry of International Trade and Industry (MITI), quoted in Borrus, Millstein, and Zysman, "International Competition," pp. 5-6.
66. Borrus, Millstein, and Zysman, "International Competition," p. 80.
67. Ibid., pp. 83-84.

68. Ibid., pp. 98–100.
69. Ibid., p. 83.
70. Ibid., p. 86.
71. Ibid., p. 89.
72. Ibid., pp. 90–91.
73. William J. Perry, Undersecretary of Defense for Research and Engineering, testimony before the U.S. Congress, Senate Committee on Armed Services, "Department of Defense Authorization for Appropriations for the Fiscal Year 1980, Part 5," hearings, p. 2,292.
74. Baldwin, *Case Studies*, p. 94.
75. *Electronics Times*, December 14, 1978, quoted in Kaldor, *Baroque Arsenal*, p. 94.
76. U.S. Department of Defense, Headquarters Air Force Systems Command, "Payoff '70," Executive Report, Manufacturing Technology Investment Strategy, Andrews Air Force Base (Washington, D.C.: 1980), p. 28.
77. U.S. Air Force Materials Laboratory, "ICAM Prospectus Update B," Wright Patterson Air Force Base (Dayton, Ohio, September 1979), p. 8.

6 THE PENTAGON AND THE FIRM

John E. Ullmann

Among a host of dangerous side-effects, the arms race has given rise to the military-industrial firm as a distinct form of organization differing drastically from what one usually perceives as private enterprise. This split is particularly important to nonmilitary, high-tech industries because the latter have generally been organized within firms that operated with a measure of flexibility, independence, and informality. These qualities contrast sharply with characteristics typical of military-industrial firms.

THE GREAT DIVIDE

It is difficult to generalize about a group as heterogenous as the commercial high-tech producers, and it is important not to become overly sentimental in describing them. Still, whatever the size or qualities of individual firms, when high-tech development has been successful, it has not only been at the cutting edge of science but of commercial risk-taking as well. For example, those firms that developed ever newer computer applications, hardware, and software succeeded by integrating technology with an exact understanding of customer needs—the hallmark of competent commercial development. Success in the computer field, however, has been the outstanding exception compared to the dismal record of most other U.S. industries. Even the successes, however, were insufficient in preventing a large-scale erosion of America's predominance in the field of new technology. American industry showed itself most vulnerable in relation to improvements in the organization and methods of production practiced by foreign competitors, Japan being by far the most important.

The missed opportunities and damage done in the past are no longer debated. At issue now is whether, as President Reagan's arms race accelerates and becomes the centerpiece of his administration, the expected increase in the high-tech content of weaponry will result in an ever greater and more deleterious concentration of high-tech efforts not only in military projects but in the military-industrial firms that serve as contractors. If military-industrial firms further increase their share of economic activities, will their mode of operation and organization become something of a norm in the American economy? The disproportionately large employment of technical talent by military firms is likely to define the limits of the technical capabilities of American industry even more so in the future, thus changing such capabilities rather drastically for the worse in international competition and in the domestic market. A combination of financial and technical profligacy, bloated payrolls, waste motion in their internal operations, bureaucratic rigidities, and technical concentration away from commercial products will increasingly lead to a condition where much of what is left of private industrial competence will have been thrown out of Mr. Reagan's so-called window of vulnerability.

In this chapter, we will review the characteristics of the military-industrial firm and concentrate on those areas that distinguish it from commercially oriented enterprises. An unprecedented degree of direct, centralized government control over every aspect of operations in military-industrial firms is at the core of the problem. Not only does this virtually institutionalize waste, it removes most of the managerial maneuverability, a characteristic essential to any firm involved in a developing, innovative industry. The adverse impact of this government control on high-tech development could, therefore, not be more critical. In our discussion, we will thus focus on the occupational structure and resulting labor practices typical of military-industrial firms. We will conclude the chapter by examining the kinds of products manufactured by military-industrial firms, and the profound implications these products have on industrial output generally.

THE MILITARY-INDUSTRIAL FIRM

A military-industrial firm is an organization engaged in the production of weapons or other specialized equipment and whose sole customer is the Department of Defense. It may also engage in related research, development, testing, and evaluation,[1] activities of special interest to high-tech industries. Like other military suppliers, high-tech organizations in the military sector may be independent corporate entities or units of a firm making other products as well; they may also be nonprofit organizations or a specialized unit within a university. Some high-tech firms may also act as suppliers to other government agencies, such as NASA, which requires products with physical characteristics similar to the military's. Major subcontractors engaged in specialized weapons work can also count as military-industrial firms when such work forms the greater volume of their business. This last definition would thus exclude firms supplying their usual product to the Department of Defense (e.g., food, fuel, and other standard supplies). For such firms, government business is essentially no different from what they normally provide for other businesses.

Military-industrial firms are privately owned. They maintain shareholders, they are organized in the hierarchical way typical of other American firms, and their securities are traded on the exchanges or over the counter whenever they are publicly owned. Governmental ownership of such securities tends to be minimal and is typically undertaken only for purposes of bailing out the enterprise. However, as will be noted below, the possibility—indeed the likelihood—of such a rescue happening is itself a symptom of a dependence on government that is absent in the business world in general.

As the military-industrial complex developed after beginnings in World War II, the Pentagon came to function more like a central office controlling a number of divisions, or what Seymour Melman called the "state-management."[2] Each of these "divisions," meaning individual military-industrial firms, may have a substantial degree of operational autonomy at the local level. The firms are certainly operated as individual profit centers. Nevertheless, the "central office" sets policy in a wide range of activities and maintains a substantial supervisory and controlling organization that sees to it that the central directives are obeyed. It could indeed be argued that the presence of local controllers and the rules under which military procurement takes place are a good deal more burdensome and detailed than is found in many a decentralized nonmilitary corporation or conglomerate. The system of rules is essentially contained in the *Armed Services Procurement Regulations* (ASPR).[3] A further amplification of the rules is set forth in *The Defense Procurement Handbook*,[4] which serves in part as a training manual for those involved in the procurement and control process, both from the Pentagon side and at the receiving end in military-industrial firms.

It is convenient to discuss this controlling influence in terms of what have come to be accepted as the central characteristics of the decision power of an individual enterprise: such an enterprise, if truly private, exercises a very substantial degree of control over what to make, how to make it, in what quantity, at what price, and with whose invested money.

In the United States, private business has traditionally resisted, to the maximum degree of its influence, any significant governmental controls that would encroach upon this decision power. Its sustained campaign against "regulations" is a case in point. It is therefore surprising that American business has over the years strained at many a regulatory gnat but swallowed the camel of Pentagon state-management. Let us consider the above aspects of control in turn.

What to Make

The Department of Defense (DoD) essentially decides what products it wants industry to make. While feedback from suppliers and follow-on ideas from existing contracts may make their way to the top of the procurement process, there is no record of a major weapon having been the product of an unsolicited bid. A mechanism for submitting such bids does exist, but it is regarded by the industry as the longest of long shots. In practice, orders for specialized products are put out for bids or negotiated without bids (which is very common) after having been specified in the greatest detail. Some of these details are the result of prior research contracts (awarded, in some cases, to the ultimate producer), but the decision power on every detail rests with the DoD.

This detailed control is different from the oversight exercised by the head office in a decentralized, multidivision firm. In such companies, each division typically has a product development department that works closely with divisional engineering, marketing, and production. These departments of product development also keep constant watch on the product line, identify opportunities, and give direction to product research. While divisions in large firms may be constrained by central office determinations of what business the company is supposed to be in, in these times of variegated conglomerates such specifications of product are likely to be deliberately vague.

The procedures of the Pentagon are in direct contrast to the inventive and forward-looking aspects of management in high-tech firms. Smaller high-tech companies are especially likely to have great flexibility in matters of innovation and the decision on products generally, but, small or large, the contrast with military practice could not be more fundamental and far-reaching. The Pentagon must keep its own "bottom line" in constant view when deciding which products to support. This decision process, with its lethal focus, resembles nothing so much as an old cartoon showing a cave couple, with the man holding a small, crude wheel: "That's very nice, dear," says the woman, "but how are you going to hit somebody with it?"

Here we encounter something of the fatal embrace of high tech by the Pentagon. Identifying a product and its market is generally the riskiest step in the innovative process, and it is comforting to firms engaged in military contracts that these issues need not be addressed. But with such guarantees, as we shall see, goes a substantial loss of independence, indeed a drastic change in the way business is done. The bargain is virtually a Faustian one, with Mephistopheles claiming rights to any patents generated, rather than to the souls tempted.

Those familiar with the legend may recall that in addition to other benefits, Faust was promised the return of some rather valued, albeit by then, somewhat deteriorated, youthful capabilities, which, in our present discussion, are analogous to the opportunity of again "working at the frontier of knowledge." Such arrangements may prove especially tempting to small firms that have had a successful product or two in the past but which are having trouble finding a suitable encore that will guarantee them a better chance of survival—a frequent and often very serious problem for such firms. An offer of generous help funded by America's taxpayers, together with the Pentagon's guidance as to what is to be done, and without many questions asked about costs, becomes an offer very difficult to refuse. Indeed, this is a case in which the attentions of Mephistopheles may be very actively solicited.

Ordinary business practice in firms opting for contracts with the military also changes. There is, first of all, an alteration in the way in which research and development contracts are awarded. Some product ideas, for example, may have issued out of previous contacts between the prospective contractor and the scientific branches of the armed forces where such requests typically originate. Eventually the job is put out for proposals (which are major engineering tasks in themselves) and, as will be noted, this method of proposal has had a profound effect on the employment structures of military contractors.

Technical proposals are highly detailed and must include not only the prospective contractor's understanding of the job and the proposed methods of solution, but an elaborate presentation of previous experience as well. This kind of exercise resembles the proposals universities must submit as part of their grantsmanship more closely than it does any normal sales quotation. Details of a firm's physical facilities or managerial organization that must accompany proposals of this sort, for example, go beyond what a business normally submits as part of the bidding procedure. In addition, before any company is considered eligible to submit a bid, it must be included on the list of approved bidders. For newer firms in innovative work, this creates something of a Catch-22 situation, one long familiar to university graduates looking for entry level positions in certain fields: one can't get on the list without prior experience but failure to be on the list is what keeps one from acquiring experience in the first place.

How to Make It

State-management of production methods is at least as pervasive as that of product selection. The ASPR devotes four pages to the listing of items over which the government exercises direct control. The list includes more than mere technical surveillance—one would expect such supervision in the manufacture of any specialized, technically oriented equipment: machine firms are accustomed to the presence of inspectors who follow the progress of a job. Engineering firms doing work on contract are likewise tolerant of customers' representatives who track the development of a design and its procurement. Technical specifications often chart in considerable detail the materials to be used, as well as quality, inspection, and performance standards.

The government control over military contractors, however, goes much further. In addition to coverage of a project's technical aspects, wage structure, benefits, insurance, and all other aspects of a firm's operation are likewise fair game for government evaluation and inspection.[5] But functional control appears to go even beyond the level of supervision normal to contracts on a cost-plus basis. Subcontractors are also subject to similar rules (depending on the nature of their work), though these controls do become somewhat less intrusive as one drops from large, prime contracts to subcontracts for more or less standard products. It is, however, exactly in the production of high-tech items that one is least likely to benefit from these lessened controls, with virtually no easening of controls if the job starts out as a research contract.

In What Quantity

Government also establishes the precise quantity of items to be manufactured. Production volume is normally subject to major decisions on the part of private management after assessing the market to be served, its prospects for expansion, or the possibility of opening up new markets or market segments and the question of whether to design the manufacturing system for mass production or small batch output.

No such discretion is afforded in military procurement. Unless government decides on the matter, there is no way of changing production volumes. Though it may, at times, be possible to alter the size of the market slightly, there is normally no way of effecting such change, except through political process. This is because the quantity of a military product is not rationally determinable, particularly at a time when—as we are assured with perhaps excessive confidence—the product is meant for deterrence rather than use. Overkill, which has existed for over twenty-five years, eliminates market saturation as an overall limit, and so individual items are ordered in part for the pork barrel, national machismo, and similar reasons having little to do with "need."

Furthermore, research and procurement are often under separate contract, and the latter may be shared by several suppliers. Such sharing is essentially determined by executive decision in the state-management and not by any sort of market movement. In order to determine quantities, finally, the *Defense Procurement Handbook* requires consideration of additional policies such as "maintaining the industrial base" (a concept inadequately explained), assisting small or minority business or labor-surplus areas, and other social policies in which government purchases act as levers. Some high-tech enterprises may benefit from special treatment since it is being shoehorned into the manufacture of ever more weaponry, but only at the cost of having high-tech preempted by the Defense Department and, in effect, largely conceding commercial work to others.

At What Price

One of the more tiresome clichés of our time is: "if they can send a man to the moon, why can't they . . . ?" It makes its way into all kinds of discussions of industrial change, as well as into advertising copy. It was used, for example, for the original promotion of a decaffeinated coffee, with the reassuring voice-over, "consider it done."

But actually, this is a nonsensical comparison. It ignores the crucial issue that in all commercial products there is a strong constraint on the price at which the product can be sold and on the cost at which it can be produced. On the other hand, a moon shot and the work of military-industrial firms in general are essentially free of such concerns. There was, and is, no way of settling in advance how much a trip to the moon ought to cost. That depends entirely on the political realities of how large a tab the public is willing to pick up.[6]

A high-tech firm without the benefit of any such institutionalized largesse often finds the economic constraints a central problem in product development. Most products are designed to do a known job in a new way, rather than something entirely new, and even in the latter case, some substitution, even if unsatisfactory, very likely exists. If the performance obtainable by a new technology can be shown to be better, then any extra user costs to buy and operate the new product must be set against the improvement which it provides. If the user's decision is made rationally, it is based on the incremental return that is derived from this comparison. Price and operating costs may not be the only purchasing incentives but experience indicates that in most markets they are extremely weighty ones. Unless these technical-economic constraints have been successfully met, the firm in effect does not have a viable product.

The Pentagon imposes its more munificent practices upon society primarily by the cost-plus-fixed-fee contracts that are still in widespread use in military procurement. Furthermore, for many years now, some 80 percent of all contracts have been negotiated rather than put out to competitive bidding. Even in the latter case, moreover, it frequently happens that the specifications for the job, especially in research contracts, are written after consultation with a favored prospective bidder.

In settling a price for a negotiated contract, the prospective contractor and the contracting officers are mandated to base their final agreement on reasonable costs plus a standard fixed fee for profit. Considerable leeway is, however, offered for engineering changes, which are the stuff that cost overruns are made of. Once a large amount has been spent on a project, it becomes difficult simply to abandon it; rather, more and more money is pumped into it by way of engineering changes that in many cases carry a provision for additional profit. The job thus seems to go on, in Jonathan Swift's apt phrase, "driven on equally by hope and despair."

Original cost estimates are generally regarded as polite fictions. As investigations of such rather spectacular flops as the TFX fighter (later to become the extremely problematic F-111) and the C-5A transport have shown, this is recognized by all parties. In situations where competitive bids are possible, as they were in these cases, awards are based in part on who is likely to be *least* badly in error in the cost estimation.

This acceptance of price errors and outright gouging differentiates the military-industrial firm very sharply from conventional private enterprise. The latter is a cost-minimizing and profit-maximizing entity, that pursues a variety of long-range and short-range goals. By contrast, the military-industrial firm is geared specifically toward maximizing sales or subsidies.[7] By virtue of the contract structure, there is little incentive for cost saving. Attempts have been made at various times to provide added incentives for timely and on-budget performance. For some contractors, this has meant juggling costs between various accounts in order to qualify for incentive payments. This practice is a matter of much concern to the government auditors who, in line with the general head office versus division arrangements, are a strong controlling presence for major military contractors. The drive toward subsidy maximization, in turn, favors the stockpiling of engineering talent, and as long as it can reasonably be charged to a job in-house, overemployment costs nothing extra. Having such "resources" available may, in fact, help in securing a new contract. Overemployment is thus endemic to the military-industrial firm.

The effect of such wasteful policies has been a sustained escalation in weapons costs that is rarely discussed. There is no consistent price index for weapons, but in 1980 the Defense Science Board estimated that it was increasing at a rate of 25 percent a year.[8] A 1970 study by the Senate Armed Services Committee had estimated a 15 percent annual escalation.[9] When, therefore, the military budget is presented as a certain percentage "above inflation," it means that the overall producer price index, which only covers commercial products, represents the standard by which to measure inflation. It says nothing about the physical volume of weaponry actually procured. Because the volume of weaponry per dollar spent has been falling, this has caused consternation among some military planners, the more extreme of whom regard this phenomenon as "unilateral disarmament."[10]

These problems are, of course, partially the consequence of the increasing technical complexity of weapons and their central role in the U.S. exploration of high technology. Still, the success of computers, for instance, lies precisely in the computer industry's ability to supply their product with ever greater (and more reliable) functional capacity in relation to the initial investment. We would not be paying $4.99 for a calculator if its chip had to carry a Pentagon-size price tag. Technical complexity is thus not a sufficient reason for exorbitant costs. It certainly was not for the successful, albeit increasingly Japanese, firms in the field. Part of the explanation must, therefore, be sought in the professional and organizational arrangements of military-industrial firms.

With Whose Money

Military-industrial firms obtain much of their working capital and, in many cases, a sizable portion of their facilities directly from the Department of Defense. The supply of working capital essentially derives from progress payments or other partial payments that assure the military-industrial firm of a sustainable cash flow. In the commercial sector, providing working capital has proved to be a major problem in times of high interest rates, inflation, and capital shortages, and insufficient cash-flow has been the leading cause of early mortality among high-tech firms. It is only when the activities of an established military-industrial firm are affected by losses in its general business that a serious danger of insolvency develops, such as occurred with Lockheed—an event that proved to be the signal for its well-known government bailout.

A second though less significant source of capital for military-industrial firms is the outright ownership of contractor facilities—buildings, machine tools, other processing or research equipment—by the Department of Defense. Here is yet another opportunity for stockpiling and overcapitalization because capital is derived directly from government funds, rather than being charged to the government when the military-industrial firm charges its usual depreciation expense. Outright DoD ownership of equipment has fluctuated due to periods when many plants and machines were sold to the contractors, usually at fire-sale prices. Many contractor-owned facilities, especially the older ones, were acquired almost free of charge from the government.

Department of Defense largesse along these lines has also led to the early introduction of numerically controlled machine tools, long before they were economically feasible in commercial work. But this case of "spinoff" turned out to be something like fool's gold. For a long period, military contractors were just about the only major market in the United States. Initially protected in this specialized market from the need to make cost-effectiveness the first criterion of design, the U.S. producers eventually found that ultimate success in making these machines reliable, reasonably priced, and available with prompt delivery lay mainly with the Japanese and West Germans. The American firms were thus quickly bested in international competition and even in the U.S. domestic market.

OCCUPATIONAL STRUCTURE AND LABOR PRACTICES

The structure of the military-industrial firm and its business environment has necessarily led to changes in the way it organizes itself and does its job. One of the most pervasive changes involves the very nature of the engineering profession.

Throughout modern times, engineers have seen their profession grow increasingly fragmented as specialization developed. Stories, such as the one about the meeting of two aerodynamicists who, upon finding out that they were in the same profession, ask each other, "What Mach number?", derive directly from the branch of aerospace engineering. But there is more at issue than technological ultraspecialization. Administratively oriented specialties like proposal writing, technical writing, and a host of interpretive functions related to military specifications have arisen in recent years, and the result of such work is a most remarkable attenuation of technical content. The practitioners are often aware of it and are not slow in private conversation to voice their most profound dissatisfaction and frustration at not being able to utilize in any meaningful way what they had learned as engineers. The military-industrial firm is thus a prime mover in the rise in administrative overhead, which is itself one of the more alarming symptoms of economic deterioration.[11]

The ultraspecialized, heavily administrative structure, closely linked to the DoD itself, has created a strongly entrenched bureaucracy. A recently coined German word for bureaucracy, *Filzokratie*, describes the current situation well. The term is derived from the word *Filz*, literally meaning felt (cloth), but the term generally connotes a social fabric in which the threads of corporate, governmental, and individual authority, responsibility, and self-interest have been woven so tightly as to be inseparable. The word has thus come to mean a kind of greedy collusion among people whose main objective is the perpetuation of the system rather than the provision of a meaningful and efficiently created output.

A major part of this bureaucracy is occupied in enforcing the secrecy that surrounds much of the work performed by the military-industrial firms. Traditional military preoccupations with clearances, the "need to know," surveillance, locks, and guards have been joined by a new wave of hysteria over the possible leakage of industrial information. This has occurred at a time when U.S. industrial performance has suffered precisely at the hands of those ideologues and military-style managers who are now so concerned with security. The volume of what might possibly be worth stealing has probably declined as a result. Scientists working within military firms are being subjected to censorship in anything they publish, including, of late, certain unclassified material. To a large extent, this holds for all who write on matters pertaining to the business they serve, but the issue is particularly sensitive in the case of military-industrial firms. Here, the staff may lose contact with much of what is going on elsewhere, and duplication of effort is only one of a variety of adverse consequences.

Solid gains in military engineering are also often impaired by the splitting of a single professional task into a variety of separate functions. The split is not only vertical (i.e., between different technical specialties) but horizontal (i.e., within jobs) as well. Value analysis (or value engineering) forms a typical example. This has become something of a separate engineering specialty, and is related, in turn, to such concepts as producibility and reliablity assurance. Its basic precepts, "omit, combine, make cheaper," are, however, what one would suppose to be essential elements of competent product development. Such concepts were once a regular part of engineering training, until the serious inception of the arms race that began after Sputnik caused the quite rapid exclusion of such subject matter from engineering curricula in favor of more "applied science." It is now necessary to second-guess the new crop of engineers who are largely ignorant of these and other principles of commercial development. Yet it was precisely this kind of understanding that made for the true glory of American engineering. The way in which technical knowledge could be integrated with practical production know-how, product design, and marketing enabled technically oriented American industries to be the very first in the development of technically sophisticated, mass-produced commercial items.

The shift in engineering education has also led to a steady erosion of skills useful in commercially oriented research. Older engineers—whose technical skills may have grown outdated but who might have known more about commercial development—are being succeeded by a generation of scientists who know the new technology but cannot turn it into viable products. The result in part has been that U.S. firms did the original research, and foreign (particularly Japanese) firms succeeded in the marketplace. However, commercial success led them to pursue their own research as well, so that the end products were often functionally superior in addition to being priced to serve their markets. This can currently be observed in the fields of robotics and super computers.

Meanwhile, the separation of research and production that originated within the U.S. military-industrial firm has become the model for a growing number of industries that now insist on having their research paid for in a separate contract, usually or preferably by the government, rather than letting it be absorbed by product sales as part of the usual risks and rewards of business.[12]

By now, these working methods greatly affect American technical capabilities. About 25 percent of all American engineers work in such environments, as do 25, 45 and 80 percent, respectively, of mechanical, electrical, and aeronautical engineers, and 20 percent each of mathematicians and computer scientists.[12] As these proportions increase, the prospects of U.S. recovery in general and high tech in particular will be severely impaired.

PRODUCTS, PROSPECTS, AND OTHER DISASTERS

Given the above situation, it should not be surprising to learn that the history of military products is replete with stretch-outs, cost overruns, outright failures, and cancellations. The present situation thus belies another old cliché, namely that military products "have to work" and that cost considerations are only secondary, whereas in civilian products, the exact opposite obtains. Whatever truth there may once have been to this belief, it has surely become largely nonsense. Many weapons simply do not work or are so expensive to produce that their small number reduces their total military value. As to the cost of military products being "secondary," what is surely meant is that so far the public has been willing to absorb gigantic expenditures. As to commercial products, their price is indeed subject to competitive pressures, but many products are in fact of very good quality and last practically forever.

The tradition of mistaken promises established by the C-5A and TFX aircraft and other early flops continued with the Main Battle Tank, and Aegis cruiser, the A-6, the F-14 and especially the F-18 aircraft, and the Hummer vehicle.[13] The last was designed to replace the jeep, a vehicle that had proved its usefulness over decades and—together with a group of foreign imitations—became indispensable to transportation in much of the world. The contagion of technical incompetence in the United States could hardly have a more painful illustration.

Sorry practices such as these, repeated over and over again, led Deputy Secretary of Defense W. Paul Thayer to tell a conference of military contractors in June, 1983, that they could cut 10 to 30 percent from their costs if they made weapons and equipment right the first time. In terms of a 1984 procurement budget of $94.1 billion, this would mean a saving of $9.4 to $28.2 billion.[14] The prospect of such large-scale waste is daunting indeed, given a projected steady rise in procurement as a proportion of total military spending. Others at the same conference complained that the United States had failed to produce high-quality goods, whether intended for the armed forces or the commercial market. One speaker noted that many industries had accepted as normal a 15 percent scrap rate for their products compared with 1 percent in Japan for comparable products.[15] For some complicated electronic products, quality was so poor that the annual cost of maintenance was more than the cost of the equipment itself. The reasons for these dismal statistics seem to be a combination of bad design, poor raw materials, inadequately trained workers, and, quite crucially, indifferent managements. Similar strong criticism has since come from General John A. Wickham, Jr., the newly appointed Army Chief of Staff.[16] In November 1983, Congress passed a law requiring a rather limited warranty by weapons producers. Even this was vigorously attacked by the Pentagon, whose sustained efforts to repeal or render that law incited a nascent constitutional confrontation over the Administration's allegedly lackluster enforcement.[17]

There is some indication that the managements of multidivision firms (in which one or two sections are engaged in military work of the kind we discuss in this chapter) recognize the lesson of the U.S. Synthetic Fuels corporation.[18] In corporate councils, such divisions are often treated very much as separate entities, and their relationship with the Pentagon is often substantially closer on a day-to-day basis than with their own head office. Those of us who have been concerned over the years with facilitating the conversion of military-oriented occupations into commercially useful ones have often been told that top managements of large firms view such transfers within their own organizations with great reluctance; their preference would be simply to sell off the divisions rather than, let us say, letting them provide extra capacity for commercial activity. This extends still further the difficult problem of conversion, which must already take into account both the refusal of the Pentagon to participate in advance planning and the lack of concern by all administrations since the first attempt to assist the process by Senator George McGovern in 1963. Overspecialization and poor work habits are bound to continue spiraling as more high tech enters into weaponry, and with all this, conversion will grow likewise more difficult. However, as administration reactions to the most recent efforts indicate, nothing is likely to be done.[19]

As noted at the outset of this chapter, commercial high-tech industries in the United States are already facing serious competition, so much, in fact, that future participation in several product areas has become quite clouded. In more specific terms, future issues, with respect to the organizational problems of military-industrial firms discussed in this chapter, may be summarized as follows:

1. The prospect of $200 billion annual deficits resulting from the arms race means that high interest rates in real terms will continue, with prospects of a national debt twice the present level and interest costs of more than $200 billion a year.
2. The finances of new firms will be especially affected. Who would wish to invest in risky enterprises if safe investments at high interest rates are available? This situation has acted as a demonstrable brake on investments for a considerable period. President Reagan's arms race will make matters much worse.
3. Fragmentation of the science of engineering will continue and probably accelerate, propelled by the kinds of employment opportunities offered by military-industrial firms. In a few areas, this will produce an upward auction in the price of particular technical talent. Meanwhile, other vital areas of machine design, metalworking, and other basic industries will decline and eventually cease to be significant internationally and, eventually, domestically as well.
4. As more and more high-tech products are used for military purposes, they will also become more specialized and thus less suitable elsewhere in the economy. A particular danger is posed by

the profligate use of high tech in weaponry where it really fulfills no other purpose than getting an otherwise outmoded piece of hardware into "the electronic age." Let it be remembered that as late as World War I, "research" of sorts was done on cavalry swords.

5. Finally, and perhaps most crucially, managements in high-tech industries will be increasingly tempted to seek the kind of sustenance that military work can offer in much the same way that the avoidance of financial risk by individual investors is most possible when safe interest rates are high. However, as the military establishment encompasses high tech, the true price of the takeover by the Pentagon state-management will make itself felt. Innovation, the very frontiers of knowledge we mentioned earlier, will be defined predominantly in military terms. For the rest of human endeavor, there will be a decline in innovation, a loss of capability in commercial development, and the kind of eclipse that all too many American manufacturing industries have already experienced. In short, the support will have been dearly bought; the future of American high-tech industries may then be best expressed by the old Italian proverb: *Chi ha pane, non ha denti*—he who has bread, has no teeth.

NOTES TO CHAPTER 6

1. Two principal analyses of the military-industrial firm are J.F. Gorgol and I. Kleinfeld, *The Military Industrial Firm* (New York: Praeger, 1972), and S. Melman, *Pentagon Capitalism* (New York: McGraw-Hill, 1970); for an early discussion of the adverse consequences to other industries, see also J.E. Ullmann, "Conversion and the Import Problem: A Confluence of Opportunities," *IEEE Spectrum* (April 1970): 55 ff.
2. Melman, *Pentagon Capitalism*, ch. 1.
3. U.S. Department of Defense, *Armed Services Procurement Regulations (ASPR)*, vol. 32 of Code of Federal Regulations (CFR) (Washington, D.C.: Government Printing Office, 1982).
4. U.S. Department of Defense, *Defense Procurement Handbook* (Washington, D.C.: Government Printing Office, 1982).
5. *ASPR*, vol. 32, vol. 1, pp. 1:80–1:84.
6. In this connection NASA has had considerable difficulty in finding an encore after it put a man on the moon. Interplanetary satellite missions have not generated anything like the same popular interest. The space shuttle has turned out to be every bit as military as the critics of the space

program had maintained all along, and the horrendously expensive "star wars" proposals are only very much more of the same, especially in the preemption of high tech by the military.

7. Melman, "Ten Propositions on the War Economy," *American Economic Review* (May 1972): 312–318. For more details, see N. Finger, *The Impact of Government Subsidies on Industrial Management*, (New York: Praeger, 1971): ch. 1.
8. Cited in A.T. Marlin, ". . . Would Drag Down the Economy," *Washington Star* (January 21, 1981): 15.
9. Cited in "Stopping the Incredible Rise in Weapons Cost," *Business Week* (February 19, 1972): 60–61.
10. C. Mohr, "Drop in U.S. Arms Spurs Debate on Military Policy," *The New York Times* (October 24, 1982): A–1.
11. J.E. Ullmann, "White-Collar Productivity and the Growth of Administrative Overhead," *National Productivity Review* (Summer 1982): 290 ff.
12. National Science Foundation, *Characteristics of Experienced Scientists and Engineers* (Washington, D.C.: Government Printing Office, 1978), Table B–13.
13. "Army's Replacement for Jeep Failing Pentagon Battle Test," *The New York Times* (May 5, 1983): A–15.
14. R. Halloran, "Pentagon Aide Says Shoddy Work Adds 10 to 30% of Military Costs," *The New York Times* (June 2, 1983): A–19.
15. Ibid.
16. R. Halloran, "Chief of Army Assails Industry on Arms Flaws," *The New York Times* (August 9, 1983): A–23.
17. "Defense's Sneak Attack on Warranty Law," *Business Week* (February 20, 1984): 30.
18. The downward auction in product quality and managerial competence shows itself directly in government-sponsored nonmilitary projects in advanced technology. The synthetic fuels effort, begun in 1980, is a prime example. The federal corporation was meant to produce oil from shale, coal, and other substances, as a way of lessening U.S. dependence on foreign oil. The corporation was set up with the modest promise that it would offer only loan guarantees. When technical, economic, and environmental hurdles looked insurmountable to private firms interested in synfuels, the synfuel corporation started to offer price guarantees. Fictitious cost estimates were followed by cost overruns, pork barrels, and assorted political favors. Worse yet, synfuels would have placed at unfair disadvantage renewable energy alternatives, none of which are politically acceptable to the Reagan administration.
19. U.S. House of Representatives, Subcommittee on Employment Opportunities Hearings on H.R. 6815, *The Defense Production Adjustment Act*, October 12, 1983 (Washington, D.C.: Government Printing Office, 1983). See, inter alia, a presentation by the present author, pp. 69 ff.

7 UNIVERSITY RESEARCH, INDUSTRIAL INNOVATION, AND THE PENTAGON

Lloyd J. Dumas

INTRODUCTION

It is hardly necessary to argue that the progress of technology has enormous impact on the character of human society for we see the effects of that progress all around us. It has shaped the context of our lives. The physical and sociological environment in which we live and work has been profoundly influenced by the technologies we employ. This influence extends to our long-term prospects for development—even for our survival as a species.

It is important to remember, however, that the particular technological pathways we follow, both in research and in its application, are a matter of social choice and *not* one of scientific necessity. Technology is not a single-lane road that we must follow; rather, it is a complex, interconnected network of roads offering a wide variety of possible paths that lead into the unknown. It is not the imperative of science, but society's choice that determines the roads traversed and those left unexplored. Given the huge effects of technology on society, this choice is an extremely powerful one.

In modern, market-oriented economies, society directs the progress of technology in two main ways. Private research by industries servicing ordinary consumer and producer markets responds to the incentives provided by the marketplace. New products or the betterment of existing products that "sell" will be favored, as will innovations that lower the cost of production. On the other hand, governmental intervention—through regulation, grants and subsidies, and its own purchases of goods and services—also bends and shapes technological development. Governmental intervention as a social force on the progress of technology will form the focus of this chapter.

THE ECONOMIC AND SOCIAL ROLE OF TECHNOLOGY

Before World War I, the largest government spenders in the United States were at the state and local levels. Government, in general, was comparatively small. Since World War II, however, the federal government has become by far the biggest spender, and its size has grown dramatically. Furthermore, for the first time in U.S. history, the military was not essentially disbanded following a war. Though the levels of military spending have varied considerably, they have remained high, year after year, for more than three and a half decades. Roughly one-half of federal "discretionary" expenditures (those not provided for by special trust funds) is today spent on the military.

According to the National Science Foundation, in the early 1980s, the United States devoted some 30 percent of its national research and development (R&D) expenditures to military and space programs. This figure was two to three times the average percentage of national R&D devoted to these noncivilian activities by other members of the Organization for Economic Development and Cooperation (basically, Western Europe, Canada, Japan, Australia, and New Zealand).[1] In the United States, over the entire decade of the 1970s, the fraction of yearly federal budget obligations for R&D going to the military and space programs averaged more than two-thirds of the total. Fifty percent of the total federal R&D obligations went to the military alone.[2]

By the mid-1970s (the most recent data accessible at this writing), 77 percent of the National Sample of R&D engineers and scientists (excluding social scientists) who received federal support for their work received it from the Department of Defense (DoD), NASA, and the Atomic Energy Commission (AEC). Again, more than 50 percent of this support issued from the DoD alone.[3] Nearly three-quarters of the engineers and scientists employed by business and industry receiving federal support at that time received it from the same three agencies.[4]

When interpreting these and other data on military-related R&D, it is important to look beyond the Department of Defense. A significant fraction of the nation's military research and development spending is *not* included in the DoD budget. Beyond the obvious fact that much of the space program has been concerned with military activities (witness the fraction of the space shuttle's missions devoted to these purposes), the entire funding for the nuclear weapons program, for example, is located outside the Defense Department's budget. Initially located in the Atomic Energy Commission's budget, it was subsequently shifted to the budget of the Department of Energy.

Clearly, military-oriented technology has been a major preoccupation of much of the nation's research and development efforts for more than four decades. And it continues to be so. It is, therefore, appropriate to question how this particular governmentally driven social force has affected the nature of technological development at two of society's main centers of R&D activities—universities and industry. The impact the resulting choice of technological pathways has had on society will be explored as well.

THE UNIVERSITY

There has always been some disagreement as to the role of the university in society, and hence to the appropriate nature of university-based research. One view, for example, holds that a university is primarily a teaching institution, not a research center. Accordingly. research activity should be restricted—in nature and in quantity—to that which services the university's teaching function. Another view holds that the university is both a center of higher learning and a research institution on a roughly equal basis. In this view, research should not be strongly antagonistic to the teaching function, but it need not have any direct or immediate connection to teaching. Yet a third opinion holds that, more than anything else, the proper role of the university in a free society is to act as a center for open discussion and inquiry, a kind of institution in which the knowledge developed and transmitted is not subject to the support of any particular vested interest, political, economic, or otherwise. The balance or connection between teaching and research activities is not as crucial as maintaining the neutrality and independence of the university as an institution. Thus, research that compromises this independence or systematically biases the product of either the university's teaching or research functions is to be strongly avoided.

During the latter part of the 1960s, debate as to the appropriateness of performing military-oriented research and development at universities raged on campuses across the country. Primarily as a result of growing opposition to U.S. involvement in Vietnam, pressure grew for universities to divorce themselves from R&D that was supportive of that war in particular, and of war-making capability in general. It was argued that the university, as an institution, had a moral responsibility to house only activities, be they teaching or research, that fostered the improvement and enhancement of life and human welfare, not its destruction. Military-oriented R&D thus had no place at a university.

Opponents of this view held that the university as an institution had no right to exclude any teaching or research activities by imposing the values of any one group. The appropriateness of a particular piece of research was a matter for the individual faculty member or student to decide. And, in any case, the university, as part of society, had an obligation to respond to the needs of that society as reflected in the priorities established by its freely elected representatives. The exclusion of military-oriented research and development from university campuses was thus not merely unpatriotic but antidemocratic.

It seems clear that no institution and no individual in a society can or should be exempted from moral responsibility for the activities they encourage or directly carry out. This is as true for businesses, nonprofit organizations, government agencies, and their employees as it is for universities and the individuals who comprise them. In addition, it is the very essence of a free society to allow for a wide variety of concepts of morality. Nevertheless, there is one aspect of military-oriented R&D that *is* obviously inconsistent with the university's special role in society. That aspect is secrecy.

If the university's contribution to a free society is to serve as a center for open discussion and inquiry, it is quite clear that secret research of any kind—whether military-oriented or for that matter business-oriented—has no place in that institution. One may choose not to discuss openly all aspects of one's research as it is proceeding in order to avoid prematurity, misinterpretation of the implications of work still in its preliminary stages, or even plagiarism or prepublication by other researchers. Such restraints are, however, very temporary and a matter of individual judgment. Work that cannot be openly discussed because of its very nature and purpose, or because of constraints external to the university (let alone the researcher), is an entirely different matter.

If the central role of the university is to teach, covert R&D activity may properly be excluded. In such a concept, research is viewed as appropriate only insofar as it enhances the basic function of teaching. Research that cannot be openly discussed can clearly not be openly taught. But secret research may be considered inappropriate on these grounds even if teaching and research are considered of equal importance. Secret research is not merely unsupportive of the teaching function, it is directly antagonistic to it.

It is an ideal of our society that no one who is academically qualified and economically capable of attending any university should be excluded from doing so. Within the university, any course offered is to be open to anyone possessing the proper prerequisite coursework. As a result, information related to secret research cannot be taught in the classroom. To the extent that a faculty member engages in research of this kind, the instructor must not only avoid speaking directly about the research itself while teaching but also must avoid saying anything that might be interpreted as closely related to that work. Otherwise, the faculty member might well run afoul of security regulations and therefore be exposed to severe penalties. Even having to worry about such a possibility cannot help but hamper teaching effectiveness, especially in more advanced courses that are likely to draw on a faculty member's special expertise.

There are those who have argued that secrecy even subverts matters pertaining to military R&D, that it may be a hindrance rather than a help to national security.[5] Among them, interestingly enough, is Edward Teller, hardly a critic of the military or a continued arms race. Teller, for example, has written:

> Science thrives on openness—researchers should, and often must, share their findings. . . . Rapid progress cannot be reconciled with central control and secrecy. The limitations we impose on ourselves by restricting information are far greater than any advantage others could gain by copying our ideas.

> . . . Adopting a policy of openness . . . would strengthen our relationships with our allies as well as illustrate the advantages of freedom to our Soviet colleagues.[6]

Under present circumstances it is quite clear that the world of military R&D is very much a world of secrecy. On this ground alone, its presence on university campuses seems broadly inconsistent with the role of universities in society.

Nevertheless, there is a great deal of military R&D performed at or administered by universities across the United States. Although direct data on the extent of this type of university R&D activity are not readily available, it is not difficult to produce a rough estimate. As of fiscal year 1980, total federal obligations for research and development were an estimated $31.9 billion, some $4.2 billion of which was allocated to universities and colleges directly.[7] An additional $2.2 billion went to federally funded research and development centers (FFRDCs) administered by universities and colleges.[8] Roughly 24 percent of federal R&D obligations to universities and colleges came from the Department of Defense, the Department of Energy, and NASA.[9] For university-administered FFRDCs, the fraction of federal obligations channeled through these three agencies was nearly 96 percent.[10] Universities and colleges thus received, directly or indirectly, roughly $3.1 billion in federal R&D obligations in FY 1980 from DoD, DOE, and NASA.[11]

An estimated $8.2 billion of R&D (funded from all sources) was performed at universities and colleges and related FFRDCs in 1980.[12] If it is assumed that the amount of federal R&D obligations is a rough estimate of federally funded research actually performed in that same year, then nearly 40 percent of R&D performed directly or indirectly under the auspices of universities and colleges was funded by DoD, DOE, and NASA. If it is arbitrarily assumed that only three-quarters of the funding from the Department of Energy and NASA could be classified as predominantly military in nature, then the military's fraction of total R&D performed at universities and colleges and the FFRDCs they administer would be slightly over 30 percent ($2.5 billion out of a total $8.2 billion).

Another approach to the same estimate is to calculate the fraction of R&D expenditures at universities and colleges funded by the federal government, and then multiply the result by the fraction of federal R&D outlays in that same year categorized as "defense-related" or "space-related." The data and results of these calculations for the years 1968, 1970, and 1972–1979 are presented in Table 7–1. These estimates of the fraction of military-related R&D performed at universities and colleges for these years are considerably higher. There are a number of possible reasons for this, including greater emphasis on basic research at universities and lesser emphasis on funding this type of research by the military-related federal agencies.[13]

Since a more comprehensive analysis is beyond the scope of this chapter, and in order to err on the side of underestimating the influence of military R&D at universities, the lower estimate derived from the first approach (30 percent) will be used. Even so, it is clear that military-related work is a major component of university research and development activity.

University faculty often find themselves under pressure to obtain external research funding, and this is particularly true of faculty in the sciences and engineering. Expensive facilities and equipment and considerable technical assistance have become increasingly prerequisite if research is to remain at the cutting edge of in these fields. Then, too, university administrators, aware that a healthy slice of outside funding can be appropriated for general "overhead" purposes, exert another kind of pressure—pressure in the form of criteria used to evaluate faculty salary increases, promotion, and tenure. These considerations can be extremely powerful for junior faculty who base an academic career on the achievement of tenure. Even tenured faculty, however, are not immune to these pressures, nor to the desire for status and recognition associated with continued funding of their own R&D projects. For these reasons and others, that multibillion dollar pot of military R&D funding constitutes a powerful magnet for faculty in pursuit of research money.

To the extent that faculty are attracted to military R&D dollars, their creative thinking may be directed along lines different from those that might otherwise be followed. Research opportunities presented by existing projects, as well as those represented by the possibility of achieving new funding, tend to be skewed to follow technological pathways of greatest interest and relevance to the military. Even researchers currently working on projects not oriented toward the military have an incentive to keep an eye open for results that might generate spinoff projects of military interest.

The big pot of military R&D dollars does more than distort faculty research. In order to be properly trained, graduate students

Table 7-1. Military- and Space-related R&D as Percent of Total R&D at Universities and Colleges.

(1) Year	(2) Total R&D Expensitures at Universities and Colleges ($ millions)	(3) Fraction of University and College R&D Expendutures Funded by Federal Government	(4) Fraction of Federal R&D Outlays "Defense-related" and "Space-related"	(5) Estimated Fraction of University and College R&D Expenditures Consisting of "Defense-related" and "Space-related" R&D Expenditures [(3) × (4)]
1968	$2,149	73.2%	79%	58%
1970	2,335	70.6	75	53
1972	2,630	68.3	73	50
1973	2,884	68.8	72	50
1974	3,023	67.2	69	46
1975	3,409	67.1	67	45
1976	3,727	67.4	67	45
1977	4,063	67.2	64	43
1978[a]	4,614	66.3	62	41
1979	5,183	66.2	61	40

a. Estimate based on data collected from doctorate-granting institutions only.

Source: National Science Foundation, *National Patterns of Science and Technology Resources, 1981* (Washington, D.C.: Government Printing Office, 1981): Column (2) from Table 51, p. 54; column (3) calculated from data in Table 51, p. 54; column (4) calculated from data in Table 13, p. 29; column (5) = column (3) × column (4).

must have research projects of their own, and they must have faculty to supervise their research as well. If the ongoing research projects at the school they are attending are heavily oriented to military purposes, and if the faculty's attention and expertise are sharply focussed in these same directions, the graduate students will also be drawn into this work. And since the nature of work performed at this formative stage of their careers will tend to define the marketability of their skills, as well as shaping their thinking, early exposure to military-oriented R&D may have a very long-term effect. This is accentuated by the fact that much of the financial aid available to graduate students in science and engineering comes in the form of research assistantships, the bulk of which (over 60 percent in the latter 1970s) are federally funded.[14]

Furthermore, since universities ordinarily consider themselves obligated to prepare their graduates to function effectively in the aspects of their chosen fields currently emphasized by society, they will tend to alter their curricula accordingly. If the highest paying, most prestigious jobs lie in military-oriented R&D, schools of engineering and science will tend to modify their course offerings and requirements to reflect this emphasis. For example, since military R&D tends to be far less cost-sensitive than civilian market-oriented R&D, courses on the cost implications of design and on the economic evaluation of technological projects will be de-emphasized. As an illustration, at the School of Engineering and Applied Science of Columbia University, there is such a course entitled "Engineering Economy." In the 1940s, it was a required course for *all* engineering undergraduates. By the 1970s, not only had it been removed from the list of requirements, but students were often discouraged from taking it by many faculty who thought it "unserious." Enrollment during the 1970s was typically on the order of 15–20 students per semester. Since military R&D requires a far greater level of specialization than is healthy for market-oriented R&D in the civilian sector, fields and subfields of study will also be more narrowly defined. Thus, even the training of students who do not enter military-oriented areas of R&D is seriously affected by the emphasis on military R&D in the society at large.

All of these problems are far more serious at the universities and colleges whose engineering and science programs are of highest repute. Military-oriented R&D funds are not evenly spread but are instead highly concentrated at a relatively small number of institutions. By way of illustration, in fiscal year 1980 the ten universities receiving the most R&D funds from the Department of Defense accounted for 56 percent of total Defense Department R&D funding to all universities and colleges. For the Department of Energy, the comparable figure was 39 percent.[15]

Because the leading institutions consider their graduates to be those who will most strongly advance the frontiers of knowledge, they are that much more sensitive to where the highest paying, most prestigious jobs are to be found, and to where society's priorities have indicated the frontiers of knowledge to be. It is at the M.I.T.'s and Stanfords of the nation where these problems tend to be most clearly seen.

Thus, the estimate that 30 percent of university-based R&D is military-oriented may be seriously understated as an indicator of the extent to which the nation's university R&D capacity is directed toward military ends. For if this 30 percent funding is claiming the talents of a much larger share of our "best and brightest" university-based scientists and engineers, the real impact on the direction of technological progress in the United States will be far greater.

INDUSTRY

Industry accounts for the largest share of the nation's research and development activity. In 1981, an estimated national total of $69.1 billion was spent on research and development, more than 70 percent of it ($49.2 billion) by industry. This effort dwarfed that of universities and their associated FFRDCs that spent an estimated $8.6 billion in that year, slightly more than 12 percent of the nation's total. Although industry provided the bulk of its own funding, some $15.8 billion, or nearly one-third of industry's own spending on R&D, was directly financed by the federal government.[16]

Data on funds provided to industry by the federal government specifically for military-related R&D are not readily available. There are data, however, on the distribution of undifferentiated federal R&D among major categories of industry for the years 1968–79, as presented in Table 7–2. It is interesting to note how large one industry looms in these figures. "Aircraft and Missiles" received an average of more than 50 percent of the total federal R&D funds provided to industry over the twelve years covered. According to the National

Science Foundation, "This industry also has the largest share of all its R&D supplied by the Government mainly by the National Aeronautics & Space Administration (NASA) and the Department of Defense (DoD)."[17] It is quite obvious that the bulk of federal R&D funds provided to this industry were for military-related projects.

Similarly, the combined "Electronic Components" and "Communications Equipment" industries received an average of more than 18 percent of total federal R&D funds to industry over the years for which data are given for both (1968–74). Even after averaging in the five years (1975–79) for which data for "Electronic Components" are not provided (i.e., assuming it received zero funding), results in a twelve-year average of nearly 16 percent of all federal funding for industrial R&D. Considering the military emphasis on the high technology components of these industries in the United States—particularly electronics—it is highly likely that much of the federal money provided to high-tech industries for R&D was in support of military-oriented research. In fact, according to the National Science Foundation, "The second ranking industry, both in Federal R&D dollars and in Federal share of all R&D dollars is electrical equipment and communication which also is largely funded by DoD."[18]

Thus, three industries highly important to the Pengagon—"Aircraft and Missiles," "Electronic Components," and "Communications Equipment"—alone received a (conservatively) estimated average of approximately two-thirds of total funds provided by the federal government to industry for research and development purposes over the years 1968–79.

Rough estimates of the overall fraction of industrial R&D expenditures that were oriented to military- and space-related purposes through direct federal funding for the period between 1968 and 1981 are given in Table 7–3. The approach is the same as that used in Table 7–1. For the entire fourteen-year period covered, these estimates average more than 25 percent.

As discussed earlier, channeling technological efforts along particular development pathways is not merely the result of direct funding. It is also, in a market-oriented economy, the result of what "sells." Here the federal government has again stepped in to divert even more of the nation's research and development capacity along military-oriented lines. The government has become a major force in the market through its continuing purchase of large amounts of increasingly sophisticated weapons and related systems. Direct, federal, military-

Table 7-2. Federal Funds for Industrial R&D Performance by Industry, 1968-1979.

Industry	*SIC Code*	*1968*	*1969*
Total		8,560	8,451
Food and kindred products	20	2	1
Textiles and apparel	22, 23	a.	a.
Lumber, wood products, & furniture	24, 25	0	0
Paper & allied products	26	a.	a.
Chemicals and allied products	28	199	192
Industrial chemicals	281-82, 286	171	165
Drugs and medicines	283	a.	a.
Other chemicals	284-85, 287-89	a.	a.
Petroleum refining	29	34	10
Rubber products	30	37	65
Stone, clay, & glass products	32	3	1
Primary metals	33	9	10
Ferrous metals and products	331-32, 3398, 3399	1	—
Nonferrous metals and products	333-36	8	9
Fabricated metal products	34	18	8
Machinery	35	340	260
Office, computing, and accounting machines	357	b.	b.
Other machinery, except electrical	35 Balance	c.	c.
Electrical equipment	36	2,333	2,390
Radio and TV receiving equipment	365	a.	a.
Electronic components / Communication equipment	367 / 366	1,526	1,557
Other electrical equipment	361-64, 369	a.	a.
Motor vehicles and motor vehicles equipment / Other transportation equipment	371 / 373-75, 379	374	290
Aircraft and missiles	372, 376	4,533	4,524
Professional and scientific instruments	38	234	237
Scientific & mechanical measuring instruments	381, 82	35	32
Optical, surgical, photographic & other instruments	383-87	199	205
Other manufacturing industries	21, 27, 31, 39	a.	a.
Nonmanufacturing industries	07-17, 41-67, 737 739, 807, 891	431	448

a. Not separately available but included in total.
b. Data not tabulated at this level prior to 1972.
c. Data not tabulated at this level prior to 1977.

Table 7-2. continued

1970	*1971*	*1972*	*1973*	*1974*	*1975*	*1976*	*1977*	*1978*	*1979*
7,779	7,666	8,017	8,145	8,220	8,605	9,561	10,521	11,163	12,342
3	2	1	1	1	a.	a.	a.	a.	a.
a.	1	1	1	a.	a.	a.	a.	a.	a.
0	a.	a.	a.	a.	0	0	0	0	2
a.	a.	2	a.	a.	a.	a.	a.	a.	a.
180	184	189	203	214	236	266	300	367	379
158	159	171	183	194	218	249	284	347	360
a.	a.	a.	a.	a.	a.	a.	a.	a.	a.
a.	a.	a.	a.	a.	a.	a.	a.	a.	a.
22	17	15	14	20	a.	52	76	121	142
71	69	123	146	a.	a.	a.	a.	a.	a.
11	10	14	15	14	a.	a.	a.	a.	a.
10	6	12	11	8	21	26	25	28	28
1	2	3	4	a.	3	4	4	5	5
9	4	10	7	a.	17	22	21	23	33
7	11	12	13	14	27	36	45	37	36
262	315	401	429	511	509	532	576	592	673
b.	b.	a.	a.	a.	486	509	546	552	627
c.	c.	c.	c,	c.	c.	c.	30	a.	a.
2,211	2,258	2,367	2,410	2,307	2,307	2,555	2,699	2,864	3,224
a.	a.	a.	a.	a.	a.	0	0	0	0
		125	146	184	a.	a.	a.	a.	a.
1,420	1,479								
		1,417	1,362	1,137	1,057	1,093	1,202	1,314	1,543
a.	a.	a.	a.	a.	a.	a.	a.	a.	a.
		293	385	288	318	383	438	450	628
314	309								
		26	39	47	47	a.	a.	a.	a.
4,005	3,864	3,970	3,899	4,000	4,428	4,921	5,541	5,816	6,132
194	164	161	160	167	172	163	174	183	223
20	14	13	11	10	15	15	22	26	35
174	150	148	149	157	157	148	152	157	188
a.	a.	a.	a.	a.	7	5	6	9	7
480	452	431	416	463	310	375	415	510	645

Source: National Science Foundation, *National Patterns of Science and Technology Resources, 1981* (Washington, D.C.: Government Printing Office, 1981): Table 39, p. 45.

Table 7-3. Direct Federally Funded Military- and Space-related R&D as Percent of Total Industrial R&D Activity.

(1) Year	*(2)* Total R&D Expenditures by Industry ($ millions)	*(3)* Fraction of Industry R&D Expenditures Funded by Federal Government	*(4)* Fraction of Federal R&D Outlays "Defense-related"	*(5)* Estimated Fraction of Industrial R&D Consisting of Direct Federally Funded "Defense-related" and "Space-related" R&D Expenditures [(3) × (4)]
1968	$17,429	49.1%	79%	39%
1969	18,308	46.2	78	36
1970	18,067	43.1	75	32
1971	18,320	41.2	73	30
1972	19,552	41.0	73	30
1973	21,249	38.3	72	28
1974	22,887	35.9	69	25
1975	24,187	35.6	67	24
1976	26,997	35.4	67	24
1977	29,928	35.2	64	23
1978	33,164	33.7	62	21
1979 (prelim.)	37,606	32.8	61	20
1980 (est.)	42,750	32.0	65	21
1981 (est.)	49,150	32.0	64	20

Source: National Science Foundation, *National Patterns of Science and Technology Resources, 1981* (Washington, D.C.: Government Printing Office, 1981): Column (2) from Table 1, p. 21; column (3) calculated from data in Table 1, p. 21; column (4) calculated from data in Table 13, p. 29.

oriented R&D funding to industry thus does not encompass the full influence of military demand on the use of the nation's technology-developing capability. As the National Science Board points out: "Complete data and information are not available regarding the objectives to which total R&D resources are directed; only in the case of Federal obligations are R&D resource data reported according to specific areas of national concern such as health, energy, and national defense."[19]

This is, of course, as true for universities as for industry. However, while it is reasonable to assume that universities and colleges would not use their own R&D funds (including state and local government support) to fund military-oriented research, it is not reasonable to assume that this is true for industry. Precisely because some segments of industry are focussed heavily on servicing the government demand for technology-intensive military goods and services, it is a virtual certainty that a significant amount of their internally funded R&D is directed toward the development of military-oriented technology. Thus, though our 30 percent estimate of the share of military R&D in total university-performed R&D must still be regarded as conservative, it is likely to be a more accurate reflection of the importance of military R&D to university R&D effort than our comparable 25 percent estimate is for industry. The true share of military-related R&D in industry is difficult to estimate given the unavailability of relevant data, but it is surely considerably larger than 25 percent.[20]

THE TECHNOLOGICAL "BRAIN DRAIN"

The foregoing analysis of the extent to which the military-related technological effort has exerted a claim on the nation's R&D capacity has been cast in terms of R&D expenditures. The overall National Science Foundation estimate presented earlier was that some 30 percent of research and development spending in the United States in the early 1980s has been devoted to the military and space programs. This is essentially consistent with the conservative estimates derived separately for both universities and for industry.

One might argue, however, it is not so much the money spent on research and development for military purposes as it is the claim of military R&D on the nation's most crucial technology-developing resource—its engineers and scientists—that is critical. It is not unreasonable to expect that there would be some sort of rough correspondence between the share of national R&D expenditures and the share of the nation's pool of engineers and scientists directed toward the development of military-oriented technology. Nevertheless, it is worth looking at the question of "brain drain" directly.

According to data for 1982 provided in the Defense Economic Impact Modeling System (DEIMS) of the Department of Defense, only about 14 percent of the nation's engineers and scientists working in industry are included in so-called "defense-induced" employment.[21] However, the methodology used in DEIMS excludes from the "defense-induced" category all employment related to arms exports: nuclear weapons research, design, testing, production programs (all of which are located outside the Department of Defense's budget), and the military-oriented part of the space program. Perhaps even more significantly, the methodology assumes that the percentage of the workforce in any given industry made up of engineers and scientists is the same in the military-serving part of the industry as in the civilian-oriented part. Yet it is clear that the technological intensity of the labor force in military-oriented segments of industry is far greater. It is not unknown for military-industrial operations to employ one engineer or scientist for every production worker, as, for example, in the Rockwell International's B-1 bomber plant in the late 1970s. Such ratios of technologists to production workers are unsupportable in civilian-oriented industry. Conservatively assuming a 50 percent greater intensity of technologists in the workforce of military-oriented segments of industry, and making a rough correction for all of the military-oriented activity completely excluded by the DEIMS methodology, it can be safely estimated that at least 30 percent of the nation's engineers and scientists are engaged in this form of activity.

The estimate can be approached from another angle by considering the data for full-time, equivalent R&D engineers and scientists provided by the National Science Foundation. These data are presented in Table 7-4 for the years 1970-80. If we, first, extract the data for three major military-oriented industry categories—"Aircraft and Missiles" and "Electrical Equipment," essentially as before, plus "Machinery"—ignoring all others (industries ignored would include tank production, ordinance, nuclear submarines, etc.); then, make the arbitrary but not unreasonable assumption that approximately two-thirds of the R&D scientists and engineers in the first two industries and only one-quarter of those in the third industry are engaged in military-related work; and then finally, divide the military-related total for those three industries by the January totals for employment of R&D engineers and scientists in industry as a whole, we would find that, for the period 1970–80, the fraction of the nation's R&D engineers and scientists employed by industry engaged in military-oriented work averages to just under one-third (Table 7-5).

These data in terms of engineering and scientific personnel are for industry only. The attempt to estimate even roughly the fraction of engineers and scientists in universities and nonprofit institutions so engaged is beyond the scope of this analysis. We shall simply assume that the estimate for industry is roughly representative of the use of engineering and scientific personnel as a whole. This seems at least plausible, given the dominance of industry as a performer of research and development in the United States, as well as the relatively close correspondence among all of the admittedly rough estimates of the military claim on the nation's R&D capacity we have derived.

Though it is inappropriate to rely too heavily on the accuracy of estimates so crudely developed, it would appear likely that a great many of the engineering and scientific personnel in the United States have been devoting their talents to the development of military-oriented technology. It is unlikely that the fraction would be substantially less than 30 percent. In all probability, it is far higher, since the calculations referred to above are conservative estimates. It is important to remember that the preemption of technological resources has been maintained at this magnitude for two or three decades of more.

Now, the kind of new technological knowledge that will ultimately emerge from any given research or development project is not wholly predictable in advance. By definition, the researchers are engaged in a quest for new knowledge, and such exploration of the unknown and untried must always involve uncertainty. However, while somewhat undeterminable, the kind of new technical knowledge developed is strongly conditioned by the nature of the problems being studied and the type of solutions being sought. Since 38 percent or more of America's engineers and scientists have been seeking military-oriented solutions to military-oriented problems for the past several decades, it should be no surprise that the development of

Table 7-4. Full-time-equivalent Number of R&D Scientists and Engineers by Industry, 1970-1980.

Industry	*SIC Code*	*January* 1970	1971
Total (January)		384.2	367.0
Food and kindred products	20	6.3	6.6
Textiles and apparel	22, 23	2.9	1.8
Lumber, wood products, & furniture	24, 25	1.2	1.8
Paper and allied products	26	5.0	5.0
Chemicals and allied products	28	40.1	42.7
Industrial chemicals	281-82, 286	21.5	21.8
Drugs and medicines	283	11.8	12.3
Other chemicals	284-85, 287-89	6.8	8.6
Petroleum refining	29	9.9	9.2
Rubber products	30	7.4	6.7
Stone, clay and glass products	32	4.6	4.3
Primary metals	33	6.5	6.6
Ferrous metals and products	331-32, 3398, 3399	3.2	3.4
Nonferrous metals and products	333-36	3.3	3.2
Fabricated metal products	34	5.9	7.1
Machinery	35	42.3	42.7
Office, computing & accounting machines	357	a.	a.
Other machinery, except electrical	35 (Balance)	b.	b.
Electrical equipment	36	100.6	91.8
Radio and TV receiving equipment	365	1.9	2.4
Electronic components	367		
		64.8	60.3
Communication equipment	366		
Other electrical equipment	361-64, 369	33.9	29.1
Motor vehicles and motor vehicles equipment	371		
		25.5	28.2
Other transportation equipment	373-75, 379		
Aircraft and missiles	372, 376	92.2	78.2
Professional and scientific instruments	38	15.0	15.1
Scientific & mechanical measuring instruments	381, 82	4.1	4.6
Optical, surgical, photographic, & other instruments	383-87	10.9	10.5
Other manufacturing industries	21, 27, 31, 39	2.6	3.8
		19.2	
Nonmanufacturing industries	07-17, 41-67, 737		
	739, 807, 891	16.3	15.6

a. Data not tabulated at this level prior to 1972.
b. Data not tabulated at this level prior to 1977.
c. Not separately available but included in total.

Table 7-4. continued

January (continued)								
1972	*1973*	*1974*	*1975*	*1976*	*1977*	*1978*	*1979*	*1980*
350.2	357.7	360.0	363.3	364.4	382.8	403.7	421.0	444.5
6.5	6.6	6.4	6.8	6.9	6.9	6.9	7.8	7.5
1.8	1.9	1.8	1.8	1.8	1.7	1.7	1.8	1.8
1.8	1.9	2.1	2.3	2.1	2.1	2.2	2.2	2.2
4.9	4.9	4.9	5.0	5.2	6.3	6.6	7.2	7.6
41.0	40.9	41.8	45.2	44.4	46.4	47.9	48.3	50.9
19.1	19.1	19.1	21.1	20.1	20.6	21.5	21.6	22.1
13.1	13.0	14.0	15.6	16.6	17.8	18.9	19.7	20.7
8.8	8.8	8.7	8.5	7.8	8.0	7.4	7.0	8.1
8.3	8.2	8.2	8.4	8.6	8.9	10.0	10.7	10.7
6.7	7.5	7.7	8.4	8.6	9.1	7.9	8.0	9.2
4.1	4.2	4.5	4.5	4.6	4.5	5.1	5.2	5.0
6.4	6.0	6.4	6.3	8.1	8.4	8.1	8.2	8.4
3.4	3.2	3.3	3.3	3.9	3.9	3.7	3.7	3.6
3.0	2.8	3.1	3.0	4.2	4.5	4.4	4.5	4.8
6.6	6.7	7.3	7.4	6.8	7.1	7.3	7.5	7.7
43.7	46.3	51.0	52.8	55.7	55.3	58.2	61.0	63.3
a.	30.1	34.5	36.1	38.1	37.7	39.3	42.4	43.0
b.	b.	b.	b.	b.	17.6	18.9	c.	c.
83.6	85.4	82.6	82.6	80.3	84.1	85.7	86.6	94.7
2.1	1.4	1.3	1.0	1.1	.9	.9	c.	c.
	9.4	9.6	10.6	10.2	13.0	14.2	c.	c.
53.2								
	45.3	42.0	40.2	37.4	38.0	40.6	42.2	45.2
28.3	29.3	29.7	30.8	31.6	32.2	30.0	29.4	31.8
	28.2	27.4	26.0	25.4	28.2	30.7	32.9	34.4
29.7								
	1.7	1.8	1.9	1.7	1.9	1.9	2.0	1.8
70.8	72.1	70.6	67.5	66.9	72.0	82.0	86.4	87.3
15.2	16.3	17.5	17.9	18.8	20.5	22.2	24.0	28.4
4.7	5.3	5.6	5.9	6.7	7.2	7.9	9.0	11.4
10.5	11.0	11.9	12.0	12.1	13.3	14.3	15.0	17.0
3.6	3.6	3.7	3.7	4.2	4.5	4.6	4.8	4.3
15.7	15.3	14.4	14.9	14.6	15.3	14.7	16.4	19.3

Source: National Science Foundation, *National Patterns of Science and Technology Resources, 981* (Washington, D.C.: Government Printing Office, 1981): Table 49, p. 52.

Table 7–5. Estimated Fraction of Industry R&D Engineers and Scientists Engaged in Military-Related R&D.

(1) Year	(2) Full-Time Equivalent R&D Engineers and Scientists in Three Selected Industries[a] ($ thousands)	(3) Estimated Number of Column (2) Personnel Engaged in Military R&D[b] ($ thousands)	(4) Total Full-Time Equivalent R&D Engineers and Scientists Employed in all Industry ($ thousands)	(5) Estimated Claim on Industry R&D Engineers and Scientists by Military R&D [(3) ÷ (4)]
1970	235.1	139.8	384.2	36%
1971	212.7	124.8	367.0	34
1972	198.9	114.4	350.2	33
1973	203.8	117.1	357.7	33
1974	204.2	115.4	360.0	32
1975	202.9	113.8	363.3	31
1976	202.9	112.5	364.4	31
1977	211.4	118.4	382.8	31
1978	225.9	126.9	403.7	31
1979	234.0	131.2	421.0	31
1980	245.3	137.8	444.5	31

a. "Aircraft and Missiles," "Electrical Equipment," and "Machinery" only.

b. 25 percent of "Machinery," 67 percent of other two industries.

Source: National Science Foundation, *National Patterns of Science and Technology Resources, 1981* (Washington, D.C.: Government Printing Office, 1981): Table 49, p. 52 and calculations based on same.

military technology has proceeded at a rapid pace in the United States. Nor should it be surprising that the growth of civilian-oriented technology has been seriously retarded.

Of course, the "spinoff" or "spillover" argument is often invoked. Claiming that military-oriented technological development produces massive improvements in areas of civilian application and so contributes to civilian technological progress, this argument makes very little sense conceptually, and more to the point, is contradicted by straightforward empirical observation. While some transferability of technical knowledge between military and civilian applications would be expected (in both directions), conceptually it is difficult to see how directing attention to one area of technical research would routinely produce an *efficient* generation of knowledge pertaining to a completely different area.

On the empirical side, a 1974 report of a committee of the National Academy of Engineering stated:

> With a few exceptions the vast technology developed by Federally funded programs since World War II has not resulted in widespread "spinoffs" of secondary or additional applications of practical products, processes and services that have made an impact on the nation's economic growth, industrial productivity, employment gains and foreign trade.[22]

The seventh annual report of the National Science Board, governing body of the National Science Foundation, expressed concern over the serious erosion of U.S. predominance in science and technology. In several international comparisons, the empirical indicators behind this concern were detailed:

> The "patent balance" of the United States fell about 30% between 1966 and 1973. . . . The decline was due both to an increasing number of U.S. patents awarded to foreign countries and a decline (in 1973) in the number of foreign patents awarded to U.S. citizens. Overall, foreign patenting increased in the United States during the period by over 65%, and by 1973 represented more than 30% of all U.S. patents granted. This suggests that the number of patentable ideas of international merit has been growing at a greater rate in other countries than in the United States.[23]

Further, the report describes the relative production of a total of 492 major innovations by the United St ıtes, the United Kingdom, Japan, West Germany, and France over the twenty year period from 1953 to 1973:

> The U.S. lead . . . declined steadily from the late 1950s to the mid-1960s, falling from 82 to 55% of the innovations. The slight upturn in later years

represents a relative rather than an absolute gain, and results primarily from a decline in the proportion of innovations produced in the United Kingdom, rather than an increase in the number of U.S. innovations.[24]

More recently, the National Science Foundation has pointed to a continuation of these downtrends:

> U.S. patenting has decreased abroad as well as at home. . . . From 1966 to 1976, U.S. patenting activity abroad declined almost 30 percent in ten industrialized countries. . . . The decline in U.S. patenting abroad could be attributable to a number of factors, including . . . a relative decline in the U.S. inventive activity.[25]

The relatively poor showing of the United States is even more remarkable considering that these data do not specifically exclude military-related technology and hence are biased in favor of the United States.

It is worth noting that Japan and West Germany did quite well in these comparisons.

> Since 1963, inventors from West Germany have received the largest number of foreign-origin U.S. patents (83,220). In fact, among U.S. foreign-origin patents, West Germany was first in 11 of the 15 major product fields and second in the remaining four. . . . Japan ranks second in the total number of U.S. patents granted to foreign investors between 1963 and 1977 (61,510). Japan has the largest number of foreign patents in three product groups . . . and is second in an additional five categories. . . . Since 1970, Japan has dramatically increased its patent activity by over 100 percent in every product field except the two areas in which it already had a large concentration of patents.[26]

Not so coincidentally, these two countries spend on defense and space only about 4 percent (Japan, 1961–75) and 20 percent (West Germany, 1961–76) of overall government R&D expenditures, as opposed to a U.S. average of about 70 percent over the comparable period.[27]

Recognition of the serious retardation of civilian technological progress is also widespread in the nation's business community. In February 1976, a *Business Week* article, "The Breakdown of U.S. Innovation," opened on an ominous note: "From boardroom to research lab there is a growing sense that something has happened to U.S. innovation. . . ."[28] Apparently that "sense" continued to grow, because by July 3, 1978, a similar story had made the cover of that journal. The article, entitled "Vanishing Innovation" began: "A grim mood prevails today among industrial research managers. America's vaunted technological superiority of the 1950s and 1960s is vanishing. . . ."[29] The government also clearly recognized that a severe problem existed, as the Carter administration ordered a massive, eighteen-month long, domestic policy review of governmental influence on industrial innovation that involved twenty-eight agencies.

Given the huge amounts of money and technical personnel that have been poured into military-related research over the past several decades in the United States, the severity of the slowdown in civilian technological progress would not have occurred if the "spinoff" or "spillover" effects had been anything more than marginal.

But if the transferability of invention and innovation between the military and civilian worlds was and is actually low, then the decades-long diversion of at least 30 percent of the R&D effort in the United States to military-related work would predictably have produced precisely the sort of civilian technological deterioration that has in fact been experienced. Furthermore, what spinoff there is may sometimes carry with it other significant societal effects—however unintentional—traceable to its military origins.

If, as earlier argued, the nature of the technological pathways explored is a matter of social choice, *not* scientific necessity, then the social context in which technological development proceeds must shape that choice. The military, looked upon as a society within a society, emphasizes uniformity, obedience, and hierarchy. The proper functioning of the unit supersedes the needs of the individuals who comprise it. It is no surprise, therefore, that military-oriented projects were the historical sources of such technological innovations as interchangeable parts in the nineteenth century and automated metal-working machinery in the twentieth century.

While in no way denigrating the positive effects these occasional spinoffs may have on the economy, it is important to be aware of the values imbedded in them. The military is, of necessity, a highly authoritarian system. As such, it is at odds with the principles of personal freedom, individuality, and pursuit of enlightened self-interest—ideals endorsed by the wider body of society in the United States. To the extent that technologies that originated in service of the military bear the inherent values of that system, care must be taken when applying such technologies in the civilian sphere to avoid the subtle corruption of those very ideals central to democratic society as a whole.[30]

MILITARY SPENDING AND ECONOMIC DECAY

It is widely recognized that civilian technological progress forms the keystone of improved productivity and economic growth. As the National Science Board puts it ". . . the contribution of R&D to economic growth and productivity is positive, significant and high, such innovation is an important factor—perhaps the most important factor—in the economic growth of the United States in this century."[31]

Civilian technological progress is that which leads to improved consumer and producer products and to more efficient methods of production. Such progress contributes to greater labor and capital productivity through the development of new machinery and improved production techniques, and consequently the more efficient use of productive resources in general. Accordingly, as the technological brain drain generated by the military sector led to a deterioration in the rate of civilian technological development, productivity rates began to collapse.

From 1947 to 1967, output per hour grew at an average annual rate of 3.4 percent in the nonagricultural business sector of the United States, according to the Council of Economic Advisors. From 1967 to 1977, that average rate of growth declined sharply to 1.9 percent per year. From 1977 to 1982, productivity growth entirely disappeared, the index of output per labor hour being the same in 1982 as it was in 1977.[32] The deterioration of productivity growth has thus been accelerating.

The improvement of productivity plays a crucial role in countering inflationary pressures, for it is sustained growth in productivity that offsets the effects of rising input costs. It is not the separate cost of labor, fuels, materials, and capital that is relevant to the determination of product price, but rather the combined cost of these productive resources *per unit* of product. Thus, the rise in labor costs, for example, might be at least partially offset by substituting cheaper capital for increasingly expensive labor or by organizing production to use labor more efficiently, or both. As long as the net result is the production of more output per unit of input, rises in input costs need not be fully translated into rises in the cost per unit of product. Correspondingly, the upward "cost-push" pressures on price will be mitigated. But productivity is nothing more than a measure of output per unit of input. Hence, rising productivity permits the absorption of rising labor or fuel prices, without full reflection of these resource-cost increases in unit cost and thus in price.

The deterioration of productivity growth substantially compromises this cost-offsetting capability. In the absence of strong productivity improvement, rising costs of labor or fuels, for example, will be translated into rising product prices. As this occurs over a whole series of industries, a self-reinforcing rise in the general level of prices or "inflation" is generated.

As the price of goods produced in America rose higher and higher (and quality too often failed to meet world standards), the nation's industries became less and less competitive vis-à-vis foreign production. Overseas markets were lost and the U.S. export position weakened. Domestic markets were lost to foreign production and the U.S. import position worsened. The progressive loss of markets induced cutbacks in U.S.-based production, resulting in high unemployment rates. This problem was exacerbated by the flight of U.S.-owned production facilities to cheap labor havens abroad, as one logical response to the inability to offset higher costs in the United States. The declining competitiveness of U.S. industry, the result of decreasing productivity growth, has generated unemployment even in the face of high product demand.

Inflation at historically high levels continued to plague the American economy until the depth of the 1980–82 recession drove the country to unemployment rates averaging nearly half of those of the Great Depression. And as the subsequent recovery has proceeded, inflation has once more become a threat, despite the persistence of high levels of unemployment.

Productivity growth continues to be "the economic linchpin of the 1980s," according to the Joint Economic Commiteee of the Congress in its mid-1979 analysis of prospects for the economy. Its warning that, as *The New York Times* put it, "The average American is likely to see his standard of living drastically reduced in the 1980s unless productivity growth is accelerated" is precisely correct.[33]

At the end of 1979, I wrote:

> During the decade of the 1970s the dynamic process of deterioration which has been described here has produced unprecedented simultaneous high inflation/high unemployment. . . . For an entire decade, the inflation rate has

> averaged near 7% at the same time the unemployment rate averaged more than 6%. The economic prognosis for the coming decade is not good. If the arms race continues unabated, and we somehow manage to survive, these rates of inflation and unemployment—rates that were viewed as horrific at the beginning of the 1970s—will look like economic good times compared to what will be commonplace by the end of the 1980s.[34]

It now appears that this wild prognostication may turn out to have been a conservative estimate.

SUMMARY AND CONCLUSIONS

Beyond its year-by-year effects, the military budget has had an enormously negative long-term impact on the functioning of the U.S. economy. It has preempted at least 30 percent of the R&D effort administered by universities and colleges (as well as a similar share of industrial R&D activity and has laid claim to the talents of a third or more of the nation's pool of scientists and engineers. Through this "brain drain" effect, the persistence of high levels of military-related technological activity has produced a severe retardation of the growth rate of civilian-related technology. The retardation, in turn, has played a key role in producing a serious slowing of the nation's productivity growth, even to the vanishing point. The failure of productivity pushed industry increasingly away from traditional cost-offsetting behavior and into a cost "pass-along" mode. Input-cost increases were routinely translated into output-price increases. And as the prices of U.S.-produced goods and services rose, domestic production priced itself more and more out of foreign and domestic markets, leading to layoffs and growing unemployment in the United States.

The effects of this damage to the competitiveness of U.S. industry, wrought by more than three decades of persistently high military spending, surfaced with a vengeance in the 1970s. Collapsing productivity, high inflation, and high unemployment have been the sad legacy of our participation in the ongoing international arms race.

It is ironic that, in our blind quest for national security through the expansion of military capabilities, we have undermined the very source of our rise to international prominence and influence—the power of industry fuelled by the spectacular efficiency of technological capability. We have not increased our power by continued expansion of the destructive capability of our armed forces. We have, in fact, diminished it. We should have taken more seriously the warning that General Dwight Eisenhower gave us when he left the presidency in 1961: "The Military Establishment not productive of itself, necessarily must feed on the energy, productivity and brainpower of the country, and if it takes too much, our total strength declines." He could not have been more correct.

The policy implications of the analysis presented here are straightforward. Tinkering with the money supply, interest rates, tax policy, and the like, may be comfortingly familiar, but it will at best produce only temporary and cosmetic improvements in the economic situation. The fundamental deterioration of the economy cannot be undone without a revitalization of productivity growth, and that will not be achieved unless and until we have moved a large fraction of the engineers and scientists now performing military-related work into productive civilian activity. The question is not so much one of economic policy as of political will.

NOTES TO CHAPTER 7

1. National Science Foundation, *National Patterns of Science and Technology Resources, 1981* (Washington, D.C.: Government Printing Office, 1981), p. 14.
2. National Science Foundation, *Science Indicators: 1978* (Washington, D.C.: Government Printing Office, 1979), p. 182.
3. National Science Foundation, Surveys of Science Resources Series, *Characteristics of the National Sample of Scientists and Engineers: 1974, Part 2, Employment* (Washington, D.C.: Government Printing Office 1975), Table B-16, pp. 128-142.
4. Ibid., Table B-15, pp. 113-127.
5. Stephen H. Unger, "National Security and the Free Flow of Technical Information," Commiteee on Scientific Freedom and Responsibility, American Association for the Advancement of Science (September 1981), pp. 12, 14.
6. Edward Teller, "Secrecy: The Road to Nowhere," *Technology Review* (October 1981).
7. National Science Foundation, *National Patterns*, Table 24, p. 35.
8. National Science Foundation, Survey of Science Resources Series, *Federal Support to Universities, Colleges and Selected Nonprofit Institutions—Fiscal Year 1980* (Washington, D.C.: Government Printing Office, 1982), Table B-41, p. 148.

9. Ibid., Table B-9, p. 29.
10. Ibid., Table B-41, p. 148.
11. Ibid., Table B-9, p. 29 and Table B-41, p. 148.
12. National Science Foundation, *National Patterns* p. 21.
13. Assuming the overall average fraction of federally funded R&D that was military-related applied to federally funded R&D at universities would lead to an overestimate of university military research under these conditions.
14. National Science Foundation, *National Patterns*, Tables 63 and 64, pp. 65 and 66.
15. Calculated from data in National Science Foundation, *Federal Support to Universities*, Table B-16, pp. 41-42.
16. National Science Foundation, *National Patterns*, Table 1, p. 21.
17. National Science Foundation, *Science Indicators*, pp. 84-85. This was specifically in the context of a discussion referring to the years 1967 and 1977.
18. Ibid., p. 85.
19. Ibid., p. 43.
20. It is also worth noting that for the same reasons the estimation approach used in Table 7-1 may have yielded higher estimates than appropriate for universities, the estimates in the final column of Table 7-3 may be biased downward. To the extent that federal military R&D funding is oriented mainly to applied research and development (and not to basic research), the emphasis on this type of R&D in industry may imply that a higher than average share of overall federal R&D is "defense-related" and "space-related" there. Thus, the share of federal R&D funds provided to industry for military purposes may be underestimated in Table 7-3.
21. U.S. Department of Defense, "Estimates of Industrial Employment by Occupation (Engineers & Scientists)," *Defense Economic Impact Modeling System (Occupation by Industry Model)* (Washington, D.C.: Government Printing Office, 1983).
22. National Academy of Engineering Committee on Technology Transfer and Utilization, "Technology Transfer and Utilization, Recommendations for Reducing the Emphasis and Correcting the Inbalance" (Washington, D.C.: National Academy of Engineering, 1974), p. i.
23. National Science Foundation, *Report of the National Science Board: 1975, Science Indicators: 1974* (Washington, D.C.: Government Printing Office, 1976), p. 17.
24. Ibid., p. 19.
25. National Science Foundation, *Report of the National Science Board: 1979, Science Indicators, 1978* (Washington, D.C.: Government Printing Office, 1979), pp. 20 and 21.
26. Ibid., pp. 19 and 20.

27. Ibid., pp. 146 and 147.
28. "The Breakdown of U.S. Innovation," *Business Week* (February 26, 1976).
29. "Vanishing Innovation," *Business Week* (July 3, 1978), p. 46.
30. An argument along these lines is developed more fully in David F. Noble, "The Social and Economic Consequences of the Military Influence on the Development of Industrial Technologies," in L. J. Dumas, ed., *The Political Economy of Arms Reduction: Reversing Economic Decay* (Boulder, Colorado: Westview Press, 1982).
31. National Science Foundation, *Science Indicators: 1976* (Washington, D.C.: Government Printing Office, 1976).
32. Council of Economic Advisors, "Annual Report" in *Economic Report of the President* (Washington, D.C.: Government Printing Office, 1983), Table B-40, p. 208.
33. Clyde H. Farnsworth, "Lag in Productivity Called Major Peril to Living Standard," *The New York Times* (August 13, 1979).
34. L. J. Dumas, "The Impact of the Military Budget on the Domestic Economy," *Current Research on Peace and Violence*, no. 2 (1980), p. 81.

8 THE PENTAGON AND THE SCIENTIST

Warren F. Davis

Midway in our life's journey, I went astray from the straight road and woke to find myself alone in a dark wood. How shall I say what wood that was! I never saw so drear, so rank, so arduous a wilderness! Its very memory gives a shape to fear.

Dante, *The Inferno*, Canto I

The Manhattan Project, established on August 13, 1942, to perfect the first atomic bomb, marked the beginning of an entirely new relationship between the American military and the research and engineering community. Contributing scientific experts were thenceforth known simply as "scientific personnel," their time-honored practice of publishing research in the open literature was curtailed, and they were obliged to submit to the strict standards of military secrecy. Eventually some 150,000 persons were employed by the project, but barely a dozen ever enjoyed a comprehensive overview of the project's plans and objectives.

In spite of reliable reports to the contrary, the notion persisted that Germany had already made a dangerous headstart in the direction of acquiring nuclear weapons. "We must have some countermeasure available to meet any possible threat of atomic warfare by Germany," recalls Robert Jungk, one of the few who understood the purpose of the project. "If we only had such a thing both Hitler and ourselves would be obliged to renounce the use of such a monstrosity."[1] In the end, the weapon built to *deter* Germany was instead *used* against a country that was not alleged to be in the nuclear arms race—Japan. The original intentions of those who conceived the new technology were not allowed to prevail.

Once accessible to political and military interests, the nuclear arms race acquired a momentum of its own, an enormous momentum.[a] In the years that have intervened since the Manhattan Project, all of the moral, ethical, and professional issues that it raised have been recapitulated and amplified many thousands of times over. Today's high technology professional is both the beneficiary and the victim of the project's legacy. And like the original cast of scientists, today's defense professionals voluntarily subordinate themselves to the cause; yield to the limitations imposed by military secrecy; conceive, design, and build the next generation of superweapons; forswear moral responsibility both for the diversion of resources from humanitarian needs and the consequences of their possible use; and risk reprisal should they question the wisdom of their military work. In exchange are offered a measure of prestige, financial incentives, intellectual challenge, and the patriotic gratitude of a nation. It is, in Freeman Dyson's words, the "Faustian bargain."[2]

The consequences of this "bargain" for the high-tech professional are manifold. On the one hand are the more-or-less direct and measurable effects on career stability that follow from the nature of the conduct of military work itself. On the other are the subtle emotional and psychological effects involving self, family, and friends that become manifest with a deeper comprehension of the moral, political, and military objectives, that one's work has helped to support.[b] Defense practitioners enter the workforce, as they would military service, at a young age before they are emotionally fully matured. At such an age they are much less likely to question authority or the wisdom of their own actions and choices. The advantage in

a. No individual, no matter how coveted that individual's contributions, could stand in the way. Indeed, the most sought after and respected scientist of the era, J. Robert Oppenheimer, director of the Los Alamos laboratory, was stripped of his authority and security clearance, and died an emotionally broken man for having resisted subsequent development of the "superbomb," the thermonuclear or hydrogen bomb. Oppenheimer became the nation's first and most prominent ex-defense worker.

b. To some extent this latter consequence may be one of the benefits of continued personal growth and maturity. Unfortunately, it may also be that some never achieve the degree of emotional maturity required to make such difficult reassessments of their lives. Moreover, little in the conduct of military work itself promotes such understanding.

military service is clear. There is likewise an advantage in defense work. But the employee may eventually begin to question the validity and motivation behind the work the individual has or is doing. Years of experience may bring a very different perspective, one which, because weapons represent the ability to inflict death and human suffering, carries a heavy moral burden. The result is often an increasingly reluctant but continued commitment to defense work in view of the alternatives at that point in one's life and career. The employee is torn between work that has become morally repugnant on the one hand, and the perception that change has become difficult or impossible on the other. In effect, a kind of trap is set that becomes more and more difficult to open and escape as the years go by.

In the pages ahead, we shall examine the effect of military defense employment on the participating scientist or engineer by following a path that runs approximately parallel to an actual career. That is, we will begin at the point when one is recruited into defense work and will consider the various options and tradeoffs that are likely to be encountered as one's career develops. There are two broad areas of concern. The first focuses on the professional aspects of career development, particularly long-term career stability. This will be discussed in two parts: the difficulties and consequences of remaining indefinitely in defense, and the prospects for getting out of a defense career. The second concern is with emotional and psychological effects resulting from the moral and ethical burden of defense work. This part of the discussion will "flesh out" some of the details underlying the "trap" mentioned above. Through both of these the pervasive effect of military secrecy, which has no counterpart in civilian employment,[c] will be evident.

ENTERING DEFENSE EMPLOYMENT

The most likely path leading to military employment begins with the defense recruiter on campus. It is here that the inducements offered by the "bargain" are usually first introduced to eager eyes and ears.

c. While trade secrets, proprietary patent agreements, and other restrictions may apply to civilian employment, these are not at all comparable in either scope, duration, or legal inflexibility to military secrecy and do not have analogously profound consequences on personal and professional life.

There is no pretense of balance. Lavish brochures depict the wonderful climate, night life, recreational facilities, and cultural activities available to employees of the company. The intellectual challenge of the work is emphasized. And the new graduate is offered the special enticement of not only being paid, but paid a salary substantially greater than in the civilian sector.[d]

The possible deleterious effects of the defense choice on one's professional career are not mentioned by the recruiter. Nor is it emphasized that the main product line of the company is the production of systems designed to kill human beings with ever-increasing efficiency. Instead, the challenges are presented in such sanitized terms as opportunities to work on "force multiplication," "decision aids," "threat generation," "command, control, and communication," "target acquisition," or "electronic countermeasures." Perhaps most important, no insight is offered as to whether the fruits of such labor can plausibly enhance the security of the nation. Such a perspective requires a detailed look at the historical record, especially since 1945, and a recognition that defense work is actually an instrument of a much larger political and foreign policy picture.

It can be argued that it is not the responsibility of the defense company or its recruiters to educate the student on these matters. Very well. But it is difficult to argue that the student need not have a comprehensive understanding of the "big picture" to which the future employee will contribute. Unfortunately, and perhaps significantly, very few universities preparing young people to develop weapons of such awesome power provide any background at all on the historical and political role of technology in modern warfare and foreign policy. Given this fact, the interview with the recruiter may be the last practical opportunity to come to grips with the enormous moral responsibility that such work carries.

And, finally, the student neither enters nor leaves the interview with the defense recruiter with an appreciation that this is a business deal in which the student's services are being purchased to fulfill the

d. It is notoriously difficult to ascertain the precise factor by which compensation for defense work exceeds that for comparable work in the civilian sector. My attempts to locate relevant studies have met with claims from academic economists to defense specialists at the Joint Congressional Economic Committee that such data simply do not exist. To economists, a refrigerator and a tank are indistinguishable. It is significant that an industry that so dominates the U.S. economy is essentially invisible from the point of view of economic analysis. Nevertheless, a generally accepted rule of thumb is that defense salaries run about 15 percent higher than those for comparable nondefense employment.

economic, political, or military aspirations of others. That this may *not* be in the long run the best "deal" possible for the individual offering services is also of no concern to the recruiter or the employer represented. Yet it is here that the engineer's or scientist's career begins, a most crucial moment in the development of that career.

SECRECY

Nearly all defense work is done under strict military secrecy. Notable exceptions are initial exploratory investigations involving pure research, which are usually done in an academic setting. But even here alarming trends have emerged under the Reagan administration.[3] Secrecy requires security clearance, and security clearance requires that the employee submit to an investigation that begins with fingerprinting and completion of the PSQ (Personnel Security Questionnaire). On the PSQ one must account for places of residence and employment for up to the past fifteen years and must give details of all organizational memberships ever held and all foreign countries in which one has resided or visited. A separate, detailed account must be provided if one has ever been a member of any of over 300 organizations appearing on a list compiled by the U.S. Attorney General during the 1950s.

Without the explicit knowledge or approval of the employee, acquaintances of all kinds will be questioned in the course of the investigation and encouraged to make statements and personal judgments about one's moral character, reliability, habits, and trustworthiness. These data will be reproduced and shared among various unspecified governmental agencies, again without the explicit knowledge of the employee. There is no obligation on the part of these agencies to reveal to the employee either the information they have collected or its sources, and no systematic procedure is available to redress erroneous or prejudicial items. These data can have an important bearing on one's entire professional career.

Though clearance may have been granted at the outset, one can be "reinvestigated" at any time, with the possible outcome that clearance may be subsequently revoked. In the case of Oppenheimer, facts that had long been in the record and had not been sufficient for denial of his initial clearance were later used to rescind it. It mattered naught that he had brilliantly directed the Los Alamos Laboratory and the design and construction of the atom bomb. A contemporary example is provided by the case of a senior principal design engineer at one of the country's major defense contractors whose security clearance is also being "reinvestigated" after he publicly questioned the ethics of military work. After decades in defense he was told, "You've made certain statements in public and we want to know if you are someone who will safeguard secrets."[4] This man's entire career now hangs in the balance as he waits powerless for the other shoe to drop.

Secrecy is imposed on defense work in the name of protecting national security interests. The extent to which security is enhanced is a matter of debate, with the current trend being to expand the scope of restricted material.[5] In the past there have been significant shifts in the opposite direction, as exemplified by the sudden declassification of U.S. fusion research in the late 1950s when it was realized that the Soviet Union had a well established and advanced program comparable to ours. A major concern of U.S. scientists was the stifling effect of classification on the U.S. program.[6] We have gone on to benefit greatly from Soviet research on this subject. Indeed, one of the more promising possibilities for production of energy by controlled fusion is the TOKAMAK, a Russian acronym for a design that originated in the Soviet Union. It is much less a matter of debate that the roots of a significant share of the professionally undesirable aspects of working in defense can be traced to the imposition of secrecy. This will become clear from the following discussion.

The imposition of secrecy enables, and even necessitates, a strong vertical authority structure within the company, with two immediate consequences. The first is that management automatically assumes relatively more power and is more immune to internal criticism than in a company that is not so structured. The second is the complement of the first: the individual employee has relatively less influence and control over the activities and policies of the company. In the long run, it is the employee who is hurt.

Under secrecy, the standard test of what one is permitted to know is defined by the "need to know." This means quite literally that one cannot have access to classified information, even though it may pertain to the project on which one is working, unless it is essential for the discharge of one's own responsibilities to the project. This leads to a high degree of compartmentalization of individual responsibilities that is, from a management point of view, desirable. It is, however, undesirable from the perspective of the employee because compartmentalized responsibility also leads to technical overspecialization. Whereas in nonclassified work both the company and the employee generally benefit from a rich and continuous cross-fertilization of ideas, this is not and cannot be so to the same extent in the blinkered environment of classified work. Rather, one tends to become an expert in a narrowly defined area, while the balance of one's training suffers from disuse. Professionally, this is the kiss of death in fields in which entire technological revolutions take place on the order of every five years.

For example, consider the hypothetical case of an electrical engineer working on electronic countermeasures for combat aircraft. One of the demanding requirements in this field is for specialized electronic filters designed to match, and hence to provide jam-proof recognition of, frequency-agile radar signals. Very closely related is the technology of surface acoustic wave (SAW) devices for the generation of the required radar signals. Both disciplines come within the purview of the electronic engineer, though it is likely that major responsibility for the two areas would be assigned to different individuals within the project. The crucial difference in the classifed environment is that, to the engineer working on filter design, the latest advances in SAW technology would be as inaccessible, due to classification, as they would be to a member of the general public. (Unless, as is actually possible, SAW technology were being used also for the filtering application. This dual capability of signal generation and filtering using SAW technology illustrates how closely allied two areas can be and yet be artificially separated by the "need to know" criterion.) Conversely, the specialized techniques of filter design for this particular application would be inaccessible to the SAW expert. Whereas in a nonclassified environment both individuals might benefit from advanced knowledge of the other's specialty, this is prohibited here by the "need to know" criterion in the interest of national security. Both engineers suffer professionally because they cannot keep pace with advances occurring in related area of their own field.

EFFECTS OF OVERSPECIALIZATION

To fully appreciate the negative effect overspecialization can have on one's career, one must be aware of a number of additional factors that bear on the problem.

The first of these, mentioned earlier, is that defense salaries are generally higher than in nondefense. This means that a defense industry employee is likely to settle into a standard of living somewhat above that of a nondefense counterpart. The financial commitments and lifestyle accommodations of the higher standard mean, in turn, that defense employees tend to need to stay in defense to maintain their higher salaries. Unfortunately, it may not be possible, for reasons beyond the control of the employee, to remain in defense indefinitely.

One perhaps counterintuitive, reason for this is linked to the primary motive of the industry and, indeed, of all capitalist industry: profits. In contrast with the impression that may have been created by the promotional materials seen in the defense recruiter's office, the company is not in business to promote the emotional, cultural, intellectual, and professional well-being of the employee, though, for a time, this is possible. The ultimate purpose for which the employee is hired is to enhance the profitmaking capacity of the company for the benefit of its stockholders. But with time the employee may actually become a liability to the company. There will have been salary increases, greater responsibilities to family and community outside the workplace, concomitantly greater reluctance to relocate for the company, and the tendency toward overspecialization created by the secrecy requirement. Balanced against these on the company's profit-loss ledger is the possibility of going back to the campus to hire a new graduate to replace the seasoned employee. Yet, this seems at first not to be in the interests of the company. Why lay off an experienced, skilled employee in exchange for a less skilled and inexperienced new graduate?

The key is that the new graduate may not in fact appear to the company to be less skilled. Today's extraordinary pace of technological progress (with which the new graduate, having just emerged from an academic environment, will be current), coupled with the overspecializing and narrowing effect of secrecy on the already-practicing employee may make the new graduate more attractive. The ability of defense companies to service current contracts and to attract new ones depends critically on their ability to exploit the very latest technological advances. It is here that the new graduate enjoys a special advantage. Indeed, many technological breakthroughs and advances are the product of research done on the campus, the environment from which the new graduate comes. Moreover, the new graduate is attractive for other tangible reasons: an initially lower salary, the willingness (and the lower cost) to relocate to suit the company, and the special enthusiasm and devotion often brought to the first job. The net effect is a tendency for defense companies to hire through one door while they fire through the other. In fact, some defense companies have become notorious for this practice.

A second way in which the employee is disadvantaged by over-specialization, and one that also makes staying in defense for an indefinite period difficult, is related to the nature of defense procurement.

Defense markets for applied technology and basic research alike are almost entirely artificial. They are usually based more on the perceived need to "send messages of resolve" to the enemy or the financial interests of defense contractors than on demonstrable military need. For example, in 1982 the Reagan administration reauthorized development and production of the controversial B-1 bomber. During the Carter administration, the B-1 was cancelled because its military usefulness as a penetrating "interdictive" bomber had been obviated by the cruise missile. The ultimate justification for proceeding now with the B-1 and its enormous expense appears to rest on the chance finding of "targets of opportunity" within the USSR by the bomber's crew in the event of nuclear war.[e] It has been claimed recently that Soviet "look-down" radar now has the ability to track flying objects as small as the cruise missile. Thus the intended capability of the B-1 to penetrate Soviet air defenses, for whatever purpose, seems doubtful even before the first aircraft comes off the production line. The motivation for building the B-1 does not reflect a demonstrable military need in the sense that a definite, quantifiable military objective is to be served. Rather, the market for the B-1 has been created artificially by the desire to send a "message of resolve" to the Soviet Union.

e. It had first been argued that an interdictive bomber (the B-1) was required to penetrate Soviet air defenses and that the current B-52s were inadequate for this purpose. The cruise missile, however, would be capable of interdiction with a 200 kiloton nuclear warhead and, at the same time, would enhance crew safety and effectiveness because it is pilotless. The B-1 was cancelled during the Carter administration for this reason.

In a question-and-answer that was deleted from the official transcript (unclassified) of the *National Security Issues 1981 Symposium* (sponsored jointly by the USAF and MITRE Corporation and held at MITRE, Bedford, Mass., on October 13 and 14, 1981; MITRE document M82-30), Lt. Gen. Robert T. Herres stated, when pressed to justify the B-1 in light of the capabilities of the cruise missile, that the manned interdictive bomber was needed in case there might be "targets of opportunity" within the Soviet Union.

The role of defense contractors in creating their own artificial markets is also significant and can result in less than desirable consequences for U.S. foreign policy. An example is provided by the events leading to deployment of U.S. long-range cruise and Pershing II missiles in Europe.

In the early 1970s, U.S. defense contractors had, for a variety of reasons, considerable difficulty eliciting support in Washington for the cruise missile. Hans Eberhard, then director of the Armaments Directorate of the German defense ministry, has stated that "the U.S. manufacturers badly wanted a European endorsement for the cruise missile. They even offered the opportunity to cooperate in production, in hopes that certain European nations would then pressure the United States to produce it in large numbers."[7] By emphasizing the accuracy and cost-effectiveness of the missile, the U.S. corporations persuaded the Europeans, who then put pressure on the Carter administration for deployment. Similarly, early in 1978 it was leaked to officials of the Martin Marietta Corporation, principal contractor for the Pershing I-A, that long-range weapons were being emphasized at the secret High Level Group (HLG) meetings in Europe. Martin Marietta responded by proposing, to officials of the U.S. Army and members of the HLG, to add a second stage to the Pershing II, which they were then developing as an upgrade of the Pershing I-A, but with the same limited range. The ability to hit targets in the Soviet Union from bases in Germany, and the apparently noncontroversial nature of the addition of a second stage to increase its range, persuaded the HLG to pressure for deployment of the long-range version of the Pershing II.[8] In both cases, it is considered "business as usual" that corporations enter directly into such negotiations, creating their own markets where none existed previously. That they have direct access to officials of the U.S. armed services and to NATO ministers is not regarded as unethical, or possibly compromising the broader security interests of the American people. In this particular example, the result has been a serious test of the cohesion of the NATO alliance and withdrawal of the Soviets from arms limitation talks. Evidently, it is acceptable that corporate interests, reflected in the creation of new military markets, prevail at our peril.

Historically, perceived military needs swing back and forth like a pendulum as administrations come and go and popular sentiment shifts. The number of jobs available in the defense sector has exhibited a corresponding cyclic pattern, alternating between boom and bust. Under the Reagan administration, defense spending has increased sharply, with the effect that it is difficult for those new to the professions to imagine that a severe reversal could occur in the near future. Yet, if the past is any measure, this is exactly what will happen. For example, in the cycle that ended in the early 1970s, some 17,000 engineers and scientists were laid off in the state of Massachusetts alone, and about 100,000 nationally.[9]

It is during the "bust" phases that overspecialization becomes a special liability. With fewer jobs of all kinds available in defense, the probability that one's particular specialization is required in sufficient numbers to maintain employment becomes small. The possibility of quickly shifting to another area of specialization is essentially nonexistent because the available positions go first to those already qualified by experience. In the 1970s bust, one of two choices was available to those laid off: either accept a temporary job in a completely unrelated area and hope for a lucky break, or leave the defense field entirely for one more secure. Few of those affected could be absorbed into technical positions in the civilian sector, though organizations sprang up to find, and even create, such opportunities.[f] People with Ph.D.'s in engineering and science took jobs in book stores or driving taxis. Others who could afford to do so went back to the university to get a degree in law or medicine. Yet others left the country for positions in Europe in what became known as the "negative brain drain."

Paradoxically, even during boom times overspecialization can make it difficult to stay in defense. Defense contracts are written for periods of, at most, a few years' duration. Because companies also tend to become overspecialized, there are likely to be only a few, or perhaps no, other companies competing for certain contracts.[10] Eventually, the contract on which one is working will perforce end. If it has been possible for the company in the meantime to land another contract requiring the same or closely related skills, there may be no disruption of employment. However, it is possible that such a contract is simply not available. In this case, the likelihood of being laid off is high, not because of a general downtrend in defense spending, but because there happens at that moment to be

f. One such organization that was formed in the early 1970s was the Association of Technical Professionals (ATP), in Wayland, Mass. This group was noteworthy for its comprehensive approach to the problem of unemployed professionals, including the introduction and promotion of state and federal job development legislation.

no further demand for one's particular specialization. With relatively few, or even no, other companies competing in the same area, there may be no alternative but to leave defense work, at least temporarily.

On the contrary, markets in the civilian sector tend to be based on real, tangible, long-term needs that are not subject to the vagaries of political perception. This has a readily understandable stabilizing effect on civilian employment opportunities.

On the one hand, in the civilian sector the basic requirement for one's skills exists independently of a particular company or contract. If one is laid off, there is a high probability both that a competing company exists and that it may have need of one's skills. This is not as likely on the defense side for two reasons: there may be few or no competitors, and the market, being artificial, may already be saturated with respect to a requirement for one's specific skills.

On the other hand, the reality of civilian markets is reflected in the relative stability of civilian companies, especially among the smaller ones. This also contributes to greater employment stability. Small firms dependent on Defense Department contracts are particularly vulnerable to shifts in defense procurement and R&D (research and development) policy. They tend not to have the financial resources or diversity required to "ride out" such changes and often simply go out of business.[g] This is not to say that smaller firms in the civilian sector do not also fail, but that they are less subject to a force that is beyond their ability to influence or predict.

LEAVING DEFENSE EMPLOYMENT

Juxtaposed with the difficulty of remaining indefinitely within the defense industry is the difficulty of being absorbed into the civilian sector as an ex-defense worker. Here again, many factors, including overspecialization, bear on the problem.

g. The Department of Defense (DoD) has recently become sensitive to the reluctance of smaller firms to take on DoD contracts for these reasons. Among other things, it has conducted a series of "road shows" in major cities across the country to try to convince smaller firms that they can safely take on DoD contracts. Secondly, it has introduced the so-called Carlucci initiatives that change established contract practices so that they are more attractive to smaller firms. These measures include the introduction of multiyear contracts, as well as "Pre-Planned Product Improvement" (P-cubed-I) which is intended to ensure the availability of follow-on contracts.

It has already been mentioned that salaries in defense are generally higher than for comparable positions outside defense and that defense employees tend naturally to adopt a commensurate standard of living. One of the first difficulties that an ex-defense worker may encounter is the inability of a potential civilian employer to compete with the salary standards of the defense industry. One might think that this difficulty could be readily circumvented if the employee would be willing to accept the somewhat lower available salary. There is, however, a hidden problem over which the employee has little practical control. Employers in the civilian sector are well aware of the defense versus nondefense salary differential and of the consequent pressures on ex-defense workers. Even if the ex-defense worker is willing to accept the lower salary, the civilian employer is often reluctant to hire such a person out of concern that the individual will subsequently leave to regain the higher salary available in defense. While in the employ of a nondefense firm, the ex-defense worker will likely have an opportunity to overcome some of the effects of overspecialization. Civilian sector employers see themselves as, in effect, subsidizing the retraining and updating that is required for the ex-defense worker to become valuable again in defense. They prefer that they, not another industry with which they compete for personnel, be the beneficiary of their investment.

Confounding the difficulties presented by overspecialization and the salary differential is the effect that secrecy may have on one's ability to sell oneself to a prospective nondefense employer. It is likely that the imposition of secrecy on one's prior defense work will have greatly restricted one's ability to publish in the accessible literature. In some cases, publication may be forbidden altogether. Thus, one of the most effective means available to communicate an impression of one's professional capabilities may be denied or curtailed. This can be a real disadvantage, especially if others whose right to publish has not been compromised are competing for the same position.

Secrecy, likewise, encumbers the other channels of recognition essential for professional advancement. Personal contacts and interactions at conferences, meetings, trade shows, and within professional societies must also conform with the dictates of military secrecy. Many important professional opportunities become available because others have been exposed to one's interests and capabilities through such exchanges. Indeed, it is the expressed purpose of most professional societies to encourage such interactions through meetings and conferences as a primary benefit of membership. Secrecy compromises the ability of this process to operate and thereby restricts one's options for advancement.

And not least, secrecy impedes direct communication of one's background and capabilities to a prospective employer. For example, one may be aware that one's particular background in radar signature analysis may suit one ideally for a position in medical imaging. But this is of little value if one is not at liberty to explain the details of radar signature analysis, how the two disciplines are similar, and, in particular, what contributions one has made to the field. Similarly, an ex-defense engineer who has designed automatic data encryption hardware might be hard-pressed to communicate anything meaningful to a prospective employer without simultaneously revealing classified processes to an interviewer.

Even when publication restrictions have not been imposed presumptively, the right to publish may be compromised if one's work is funded by the DoD. A now-famous example is the SPIE (Society of Photo-optical Instrumentation Engineers) conference held in San Diego in August 1982, an open, unrestricted scientific meeting. Just days before it began, the Pentagon summarily withdrew presentation of nearly 150 unclassified papers.[11] The legal basis for this action rested not on the classification of the papers but on the presence of Soviet scientists at the conference, which potentially conflicted with certain provisions of the Export Administration Regulations (EAR, Department of Commerce) and the International Traffic in Arms Regulations (ITAR, Department of State). For example, the EAR include a Military Critical Technologies List (MCTL), the current version of which amounts to a 700-page document classified as "Secret" in its entirety.[12] Hence, those without security clearance cannot examine the list to determine if they might be in violation. In the case of the SPIE conference, it was argued that the mere presence of the Soviet scientists risked export of a critical technology to the Soviet Union, thus requiring withdrawal of a significant fraction of the SPIE papers. The proximate basis for the DoD cancellations was its funding of the (unclassified) research that the papers described.

There have been repeated attempts by the DoD or the Department of State to block presentation of scheduled papers at scientific conferences; the SPIE incident is but one, rather impressive, example. In November 1982, just three months after the SPIE conference, several papers were withdrawn from the Optical Society of America meeting in Tucson, Arizona. In the three years from 1981 to 1983, there were at least nine such attempts by the Department of Defense, prompting the American Association for the Advancement of Science to undertake compilation of a complete list.[13]

These restrictions on the exchange of scientific information, whether presumptively or unexpectedly, have a detrimental effect on U.S. science and industrial technology, as well as on the defense worker personally. In fact, our entire culture is now based on science and technology. Its vitality, and its ability to be carried forward and advanced, is derived from the free exchange of information, the greatest single determinant of which is unrestricted journal publication. We do violence to ourselves and our culture when our zeal to thwart the Soviet Union reaches paranoiac proportions. A recent study of the appropriate balance between the interests of national security and scientific communication conducted under the auspices of the National Academies of Science and Engineering concluded that " . . . the best way to ensure long-term national security lies in a strategy of 'security by accomplishment,' and that an essential ingredient of technological accomplishment is open and free scientific communication."[14] And, Robert Marshak, president of The American Physical Society, has perhaps best expressed the concern of the scientific community in the following editorial observation:

> Why has the Soviet Union found it necessary to rely so heavily on Western technology? Science education in the Soviet Union is a source of envy to those of us concerned with the crisis in our own schools. Individually, Soviet scientists are as dedicated and creative as any in the world. The Soviet government is generous in its support of basic and applied science. And yet, experimental science in the Soviet Union is scandalously bad. Much of the explanation, I believe, lies in the fact that the Soviets have created barriers to free communication for their own scientists not unlike those that some would impose here.[15]

RELEVANT SKILLS

Just as there are inherent differences in the markets served, so too are there differences in the fundamental objectives and approaches to work in and outside of defense, with the result that skills acquired through years of experience in military employment may not be readily marketable in the civilian sector. Weapons systems, and their associated support systems, must be designed and built to function reliably in the extreme conditions of battle. These include extremes of temperature, vibration, shock, incident radiation, pressure, and humidity. There are no comparable requirements placed on products intended for civilian markets. No one realistically expects a commercial typewriter that has been frozen, thrown against a concrete wall, and soaked in water to function properly. Yet, military hardware must be capable of withstanding comparable abuse. To achieve such reliability is expensive indeed and would create price tags far beyond the reach of civilian markets.

By comparison, nondefense products must be designed and manufactured to minimize cost while maintaining reliable performance within the relaxed constraints of a peacetime environment. First, the philosophy of the design of the product itself is fundamentally different in the civilian sector. For example, mechanical strength might be imparted to the product through the use of folds in the sheet metal of its subassemblies, adding stiffness without substantially increasing the cost of materials or fabrication. Castings of special alloys might be required, instead, if the product were destined for a defense application, greatly increasing both the cost of materials and fabrication. Second, overhead costs must be minimized because they are ultimately passed on to the consumer. In defense applications, it is not unusual for the design objectives to necessitate acquisition of very expensive special-purpose hardware, for both the design and manufacturing phases, which may be underutilized by civilian standards. The cost, for example, of acquiring a special-purpose precision lathe that may spend most of its time idle may be readily justified because the overall design objective places reliability before cost, and the lathe is clearly required for this purpose. On the civilian side, such an acquisition and the design that necessitates it would be quite unacceptable. And third, defense products usually require, proportionately, a very large research and development investment for a relatively limited production run. This is again a consequence of the reliability-before-cost philosophy in defense, as well as the artificiality of the market being served. That is, the size of the production run is limited by bureaucratically determined factors, not by the inherent market demand for the product.

The effect of these three differences is apparent throughout the spectrum of individual talents required by a successful high technology enterprise, from the orientation of the technician and engineer to the practices of management. Civilian industry sees those whose backgrounds have been cast in defense as lacking "cost consciousness." It fears that engineers who have spent years designing for reliability-before-cost and managers who are attuned to efficiency within the compartmentalized environment of the classified facility cannot readily reorient themselves for the optimization of profits. The skills of the ex-defense worker, though highly developed, may not be especially attractive or relevant to a prospective civilian employer.

MORALITY OF WEAPONSMAKING

We all, as we go through life, acquire almost unconsciously a complex system of beliefs provided and reinforced by our culture. Of the many facets of this phenomenon, Jonathan Schell has singled out the concept of "sovereign state" as the foundation upon which the apparently inexorable arms race is built.[16] Power is concentrated in state leadership in order to guarantee state sovereignty. Those who seek power have long recognized and exploited the threat of an external enemy, in the form of another sovereign state, to manipulate and motivate popular support. So strong is the tendency to embrace this process that hundreds of millions have sacrificed their lives in its cause. It is tempting, in fact, to ascribe the willingness with which human populations fall behind leadership, forsaking, if necessary, accountability, critical understanding, and life itself, to a primitive, instinctual territorial imperative.

In this picture, the special danger of the high technology nuclear arms race is apparent in the inappropriate and increasing disparity between the scale of the sovereign states, which has remained essentially fixed for hundreds of years, and the scale of the weapons systems thrown up to "defend" them. Whereas the states have remained fixed, the range and power of weapons systems have reached global proportions. For example, missiles from the Trident submarine can strike targets within the Soviet Union from an operational area of over 40 million square miles, the greater portion of the earth's seas. J. R. Oppenheimer referred to the dangers inherent in these incommensurate scales by using the metaphor of two scorpions in a bottle, stingers raised over each other. The striking by one cannot be made without the counterstrike of the other. In the case of nuclear arms, the size of the area to be defended has not been decreased; rather, the range of the weapons has been increased, thus upsetting the balance. As in the case of the scorpions, there has been no corresponding modification of the instincts that impel behavior. Therein lies the great danger.

Defense workers are caught up in this belief system in a special way. It is they who will conceive, design, and build the weapons needed to support the imperative of the sovereign state. Defense workers possess highly specialized and indispensable skills. It is essential, therefore, that the proper environment and beliefs be maintained and reinforced to assure their willing cooperation in the enterprise.

Cooperation is assured by the belief that, by building ever more sophisticated weapons, one contributes to the "defense" of the nation. Such weapons will create "bargaining chips" with which to bring the Soviet Union to the negotiating table. We are the good guys, creating new weapons, and even whole new weapons technologies, regretfully in response to the ever present Soviet threat. We do so only to create greater "deterrence" against Soviet ambitions. But deployment of our weapons is inherently stabilizing; those of the Soviet Union destabilizing. Everyone knows that we have no sinister intentions, and neither we nor our friends have anything to fear from our massive stockpiles. The converse cannot be said of the Soviet Union.[h] If we perceive ourselves to be behind in weapons technology, then we must forge ahead with programs of "modernization" and "rearmament" because it is acceptable to deal with the Soviets only from "a position of strength." If the Soviets perceive themselves to be behind, their response, on the contrary, will be to come to the negotiating table.

Ideas such as these translate into powerful motivating forces for the high technologist who is privileged to be able to do the job, and for the taxpayer as well. Unfortunately, the prescription (according to the idyllic beliefs) and reality are worlds apart. Nowhere is this disparity more evident than in the distortion, or rather perversion, of the meanings of the words *defense* and *deterrence* themselves.

h. Even the refusal of the United States to declare a policy of no-first-use of nuclear weapons is rationalized as a regrettable measure required in the interest of peace and national security. The Soviet Union, being opposed to peace and its own national security, has gone ahead and rashly adopted a no-first-use policy.

DEFENSE AND DETERRENCE

In the early 1960s, the military establishment of the United States recognized that it was not possible to defend the civilian population against nuclear ICBM attack from the Soviet Union. This recognition was a significant factor in the subsequent attainment of the ABM treaty of 1972. Yet, after twenty years, the population as a whole remains unaware, or at best is unclear, that the definition of "defense" has changed radically. Not since the early 1960s has there been any *direct* relationship between defense and the protection of human life in our country. Rather, what goes under the rubric of "defense" is actually fielded to defend U.S. strategic weapons resources, not perfectly, but, it is argued, well enough to ensure that we can inflict grievous damage on the Soviet Union. Administrations and other officials have consistently failed to communicate this distinction in unequivocal terms to the public they serve, an ambiguity that has not been without value. After all, who, in the pre-1960s definition of the word, could be opposed to the *defense* of the country, or the recommendations of the Secretary of *Defense*, or the president's *defense* budget?

Yet technology advances relentlessly, and the already altered definition of defense must be broadened correspondingly. Today's defense technologist is concerned with the "computerized battlefield." Robotics, which is introducing pilotless aircraft, tanks, and other weapons, has been hailed as the greatest advance in combat effectiveness since the introduction of the computer.[17] Advanced cockpit electronics, which execute spoken commands in milliseconds, have been recently flight-tested.[18] Lt. Gen. Brent Scowcroft[i] has defined the "automatic phase" of a strategic nuclear conflict as the "transattack" period during which "quick response systems are discharged . . . more or less automatically."[19] The trend toward automatic, computerized warfare has been forced upon us by the ever increasing variety, speed, accuracy, and lethality of modern weapons systems. Human minds cannot sense, assimilate, and react upon information with sufficient speed to conduct warfare under these circumstances. As this process evolves, it becomes increasingly likely that the civili-

i. Scowcroft later chaired the President's Commission on Strategic Forces, the so-called Scowcroft Commission, which advised President Reagan on MX basing.

zations for whom the battle is fought will have perished even before the outcome can be known. More and more, the term defense is being extended beyond its 1960s' redefinition to encompass the defense of mindless machines intended to conduct warfare in our behalf.[j]

There are, of course, many, both in and outside of government, who do realize that defense no longer means protection of the civilian population in the direct sense of the word. For those among them who nevertheless advocate increased "defense" expenditures, a new term has crept into usage—deterrence. The logic goes that, while we may not be able to save the people directly, the ability to assure the destruction of the Soviet Union will *deter* the Soviets from attacking and thus, indirectly, will have contributed to the safety of our people.

There may perhaps have been a time early in the arms race when this argument was appropriate. But with the attainment of the ability by both superpowers utterly to destroy the other, the word deterrence, like the word defense, has had to take on another meaning. The goal now is to maximize the *perception* by the other side that any transgression could trigger an overwhelming response, including use of the nuclear option. The consequences of initiating hostile action must be so certain and so catastrophic that they will be deterred from initiating even conventional warfare. It is the *perceptions* of the other side that matter. In these terms, the significance of huge "overkill" factors for our strategic forces can be understood when otherwise their only purpose would be to "make the rubble bounce." It becomes clear why we forward-base vulnerable theater nuclear forces in Europe, threaten launch-on-warning, deploy first-strike missiles, brandish a forthcoming "preemptive" capability,[20] and even why we refuse to declare a no-first-use policy. By increasing the Soviet perception of danger, we "increase the sensitivity of the trip

j. A subtle, but most important, misappropriation of the word defense is represented by offensive research masquerading as defensive research. A typical example is a study to determine how, if they tried, the Soviets might burn out the MILSTAR communications satellite system by beaming high-power microwaves at it. The study is promoted as being in the interest of defending our military resources; in reality, the results of the study will have direct application to development of a microwave beam of our own with which to threaten Soviet satellites. Through such ruses, many who believe they are doing purely defensive research are unwittingly providing the feasibility studies required for new offensive weapons systems.

wire" and thereby, it is argued, increase the deterrent value of our forces.

The principle that underlies this theory of deterrence is none other than to *increase the risk of nuclear war.* Stripped of the official rhetoric in which they are invariably packaged, every proposal for increased deterrence comes down to this singular principle. We have long since gone beyond the only other possible basis for deterrence, total destruction of the other side. Today, to work on, or to pay for as a taxpayer, increased deterrence is synonymous with and equivalent to increasing the risk of nuclear war. In fact, if this were not the intent, it would not be "increased deterrence."

Richard Burt, U.S. Assistant Secretary of State for European Affairs, has made this policy explicit in relation to the deployment of long-range theater nuclear forces (Pershing II and cruise missiles) in Europe. Contrary to widely held public opinion, deployment of these forces is not primarily a response to the Soviet deployment of its SS-20 missiles. Rather, it is intended to meet the political objective of assuring continued U.S. commitment to the defense of Europe by increasing the likelihood of strategic nuclear war, involving the United States, should the Soviets attack Europe.[21] Burt has stated:

> Thus the emplacement of long-range U.S. cruise and ballistic missiles in Europe makes escalation of any nuclear war in Europe to an intercontinental exchange even more likely. This is why our allies asked for such a deployment. This is why the United States accepted. This is why the deployment strengthens deterrence.[22]

THE HISTORICAL RECORD

Complementing the distortions of language required to support this pervasive belief system are distortions or patent oversights of historical fact. For nearly forty years, the United States and the Soviet Union each have pursued technological military superiority over the other, but both are much less secure now than when they first entered the arms race. In that time, the United States has led the Soviet Union in weapons technology with few exceptions. But what is invariably overlooked when we make yet another new technological proposal is that the Soviet Union, too, is technologically advanced; it has repeatedly demonstrated its ability to "catch up" to every such challenge presented to it by us. And, in spite of the continuous advantage that the United States has had, it has not been possible to stop and reverse the arms race. This is due largely to the consistent failure of negotiators to take advantage of opportunities afforded them by defense technology, the so-called "bargaining chips." It also reflects the inevitable ability of each side to find technological fixes as substitutes for failed negotiations, which in turn highlights the failure of technology by itself to halt the arms race.

Another prevailing fact is that every technological fix, whether by us or the Soviets, has a technological counterfix. This is grounded in the nature of technology itself and supported fully by the historical record. When the Soviets have been rebuffed in negotiations, they have always been able to draw upon their technological base to implement the counterfix appropriate to our latest fix. We have no more power to prevent this than they should the roles be reversed. We cannot, therefore, stymie or "bring the Soviets to their knees" by any conceivable advance in weapons technology. There never was and there never will be a technological fix to a problem that is basically political in nature.

The term "bargaining chip," with which we are supposed to get the Soviets to the negotiating table, deserves special comment. It enjoys the distinction in arms race parlance of representing both a distortion of meaning and of historical fact. First, in terms of meaning, a bargaining chip is, in common usage, something one is willing to throw away in exchange for suitable concessions from the other side. But, in the context of the arms race, it retains only half its proper meaning; that is, to get the Soviets to the negotiating table. In fact, according to this limited definition, we have indeed created "bargaining chips," and they have indeed brought the Soviets to the negotiating table. But the historical record shows that no "bargaining chip" created through technological military superiority by the United States has ever been used in the fullest sense. In some cases, such as the deployment of cruise missiles in Europe, there has even been frank consternation that a ". . . chip [might be] bargained away in exchange for substantial Soviet concessions in the strategic arms limitation talks."[23]

Having gotten the Soviets to negotiate, all "bargaining chips" have been, in the final analysis, nonnegotiable. The perspective invariably develops that the "bargaining chip" represents a potential strategic advantage that we couldn't possibly throw away, especially to a country that doesn't yet challenge us in that area, and the "bargaining chip" is deployed.[k] Used in this limited way, "bargaining chips" are, more precisely, "bargaining clubs."

A typical example of a "bargaining club," alias "bargaining chip," is the MX missile. The Scowcroft Commission recommended deploying 100 MX missiles in Minuteman silos as a "bargaining chip" to get the Soviet Union to negotiate and behave in a way acceptable to the U.S. administration. More missiles might be deployed if required. No recommendations were made for drawing down the MX arsenal in exchange for Soviet concessions.

THE MORAL BURDEN

There is growing awareness among scientists and engineers in defense that something may be wrong with this system of beliefs, though it is certainly true that not all are equally concerned. Some simply take these ideas at face value and do not question the morality of their work. Others sense that the whole structure might not be built on the most secure foundation but are able to rationalize what they do. Decades have gone by during which they have become accustomed to pushing the latest, most advanced weapons system out the door, never to be concerned with it again. Time and experience have taught them that producing such systems is demonstrably safe. "No need to worry." "It will never be used anyway." "You just take the money and run."

k. A stunning example is the decision "to MIRV" (Multiple Independently-targeted Reentry Vehicle) our strategic missiles in the early 1970s, which, incidentally, led to the "window of vulnerability" of the 1980s. At the SALT talks in Vienna in May 1970, the Soviet Union made twenty-six separate overtures to the United States, which by then had developed MIRV capability, to enter into agreements to limit MIRV deployment. The response of the Nixon administration was to refuse to regard the twenty-six contacts as serious, saying that it was "a trick designed to dupe the United States into delaying MIRV deployment."[24] Similar situations exist currently with respect to cruise missiles and anti-satellite weapons.

An even more remarkable example is the withdrawal on September 6, 1955, of United States support for a UN Disarmament Commission proposal that would have totally ended the nuclear arms race. The proposal, which had been introduced jointly by Britain and France, had the strong support of the United States and all other Western delegates to the commission. Only the Soviet Union remained in opposition. When the Soviet Union countered by introducing its own proposal, which contained all the main features of the British-French proposal, the United States withdrew support without explanation.[25]

Unfortunately, and perhaps not by accident, it is not easy to break out of this pattern. As important as these questions are, they are simply not discussed in the workplace. To do so would challenge the common system of beliefs, something not easy to be the first to do, and would risk the label "unpatriotic," or possibly "soft on Communism." Moreover, secrecy and security clearance place an effective damper on any impulse to ask questions or raise issues of morality. There is the ever present possibility that one's clearance may be "reinvestigated" and one's job subsequently lost.[26] Though in many ways "closest to the action," the defense worker is effectively isolated by this process. If there is any tendency to want not to face the broader implications of one's work, this is the ideal environment.

There is, nevertheless, a third group, whose numbers are increasing, that has begun to examine the assumptions of the belief system in detail. For those who finally realize that they are defending weapons and not people, that their contributions to deterrence mean increasing the risk of nuclear war, that the political process has consistently failed to deliver on the "bargaining chips" that they provide, and that the quest for technological military superiority is demonstrably futile, the emotional and psychological consequences can be profound.

Depression or withdrawal resulting from inner conflict can strongly affect relationships with family members and friends. Some experience shame and are unable to talk about their work, not because of the restrictions of military secrecy, but simply because of the guilt they feel.[1] Those within the family may have reason to counter-react as, for example, a wife who sees her husband's weapons work as antithetical to the role of the family in renewing human life. And, entirely new cross-generational conflicts that appear to be exclusively the product of the defense phenomenon are emerging. More and more mothers are expressing concern about the morality of their sons entering careers in defense, as are daughters with their fathers who already work in defense. The corresponding concerns of fathers for daughters and sons for mothers have so far not been apparent, though many women now have careers in defense. One young woman recently considering a position with a defense firm was able to justify

1. It is seldom that one is forbidden by secrecy from any discussion whatsoever of one's work. Rather, discussion of the general character of one's work is usually allowed, but specifics that would permit an enemy to undermine performance of the system being developed (frequencies of operation, detection thresholds, mechanical dimensions, etc.) are not.

her involvement to her own satisfaction, but could not face telling her future children what she had done.[m] For some, spiritual and religious concerns are uppermost. Clergy report an increasing number of defense workers turning to them for guidance, expressing clearly the conflict they feel between their obligation to provide well for their families on the one hand, and the possibility that they may be sealing the fate of those they love most on the other.[n]

This predicament is not enviable. Some anguish for years as they try to resolve the conflict they feel. Moreover, there is no easy escape. Whatever decision is made, there is inevitable uncertainty and sacrifice. Once the perspective of the belief system has been surmounted, even avoidance will not provide a satisfactory or permanent refuge. One's work and profession suffer. One can no longer go forward, but to retreat may mean leaving one's profession altogether. Each person to arrive at this crossroad must make a choice. The support of religious leaders, the community, friends, and family is invaluable, but in the end the choice is unavoidably a personal one and a painful one.

m. These observations are based on analysis of unsolicited correspondence received by High Technology Professionals for Peace (HTPFP) over a period of approximately two years.

n. In a survey in the New England area conducted by HTPFP and Social Workers for Peace and Nuclear Disarmament, approximately half of the clergy from all faiths indicated that they had been approached by defense workers expressing moral and ethical concerns with their work. Annual conferences at the Andover-Newton Theological School, Newton, Mass., have been held in 1982 and 1983 to bring together clergy and defense workers to promote greater awareness and understanding of this problem.

NOTES TO CHAPTER 8

1. Robert Jungk, *Brighter than a Thousand Suns* (New York: Harcourt Brace Jovanovich, Inc., 1958), p. 113.
2. From an interview that appears in the movie *The Day After Trinity.*
3. Adm. Bobby R. Inman, "Classifying Science: A Government Proposal," *Aviation Week & Space Technology* (February 8, 1982): 10; James R. Ferguson, "Scientific Freedom, National Security, and the First Amendment," *Science* 221 (August 12, 1983): 620-624; and Dale Gorson, "What Price Security?" *Physics Today* (February 1983): 42-47.
4. Bob Davis, "Work on Weapons Pains the Conscience of Some Engineers," *The Wall Street Journal*, July 5, 1983, p. 1. Another engineer, whose moral reservations with his past defense work were expressed in this same WSJ article, is also being "reinvestigated," apparently solely on the basis of his comments and name in the article.
5. Inman, "Classifying Science," p. 10; Ferguson, "Scientific Freedom," pp. 620-624; and Corson, "What Price Security?" pp. 42-47.
6. Joan L. Bromberg, *Fusion: Science, Politics and the Invention of a New Energy Source* (Cambridge, Mass.: The MIT Press, 1982).
7. R. Jeffrey Smith, "Missile Deployments Roil Europe," *Science* 223 (January 27, 1984): 371-376.
8. Ibid.
9. Deckle McLean, "When Professionals Lose their Jobs," *The Boston Globe Sunday Magazine* (April 16, 1972): 20; Donald White, "IEEE Report: Mostly Gloomy," *The Boston Globe* (evening edition) August 30, 1971, p. 32.
10. "U.S. Defense Spending: Are Billions Being Wasted?" *Time*, March 7, 1983, p. 26.
11. Ross Gelbspan, "When scientists get aid from U.S.," *The Boston Globe*, January 23, 1984, p. 1.
12. Corson, "What Price Security?" pp. 42-47.
13. Gelbspan, "When scientists get aid," p. 1.
14. Corson, "What Price Security?" pp. 42-47.
15. Robert E. Marshak, "The peril of curbing scientific freedom," *Physics Today* (January 1984): 192.
16. Jonathan Schell, *The Fate of the Earth* (New York: Knopf, 1982), p. 187.
17. C. J. Garvey and J. J. Richardson, "Robotics Finds Applications in Ground Warfare," *Military Electronics/Countermeasures* 9, no. 3 (March 1983): 85-91.
18. L. Farhat, "Cockpit Electronics System Reacts to Spoken Commands," *Military Electronics/Countermeasures* 9, no. 3 (March 1983): 68.

19. Address by Lt. Gen. Brent Scowcroft, *National Security Issues 1981 Symposium*, MITRE Corporation, Bedford, Mass., October 13–14, 1981. See transcript, MITRE Document M82–30, p. 94.
20. Address by Dr. Richard D. DeLauer, *National Security Issues 1981 Symposium* MITRE Corporation, Bedford, Mass., October 13–14, 1981. Dr. DeLauer's remarks were deleted from the official transcript of the symposium, though a verbatim transcript has been prepared by the author. Dr. DeLauer referred to a "preemptive" counterforce capability for the new Trident D–5 missile coming on line in 1989.
21. Smith, "Missile Deployments," pp. 371–376.
22. Richard Burt, "NATO and Nuclear Deterrence," in M. McGraw Olive and J. Porro, eds., *Nuclear Weapons in Europe: Modernization and Limitation* (Lexington, Mass.: D. C. Heath and Co., 1983), p. 111.
23. Smith, "Missile Deployments," pp. 371–376.
24. William C. Selover, "Foot-dragging at SALT?," *The Christian Science Monitor* (May 12, 1970): p. 1.
25. Steve J. Heims, *John von Neumann and Norbert Weiner: From Mathematics to the Technologies of Life and Death* (Cambridge, Mass.: The MIT Press, 1980), p. 268.
26. Davis, "Work on Weapons," p. 1.

9 LABOR, AUTOMATION, AND REGIONAL DEVELOPMENT

Tom Schlesinger

> If we are to change the rules [on productivity and government-corporate relations], and I think we must, then our timing couldn't be better. There is a national movement building for what is being called the reindustrialization of America. The recent NBC White Paper on Japan, the lead article in the 30 June *Business Week*, and the Stevenson Bill in Congress are all key indicators of a growing national concern. Peter Drucker, with his usual foresight, has called for a restoration of our international competitiveness based on three approaches: moving to more automation, the redesign of entire plants and processes are integrated flow systems, and integration of mini and microcomputers into our tools. Whether the nation moves towards Drucker's vision of the future or merely falls back to protecting historical employment opportunities may depend in some way on the results of today's seminar and the vigor with which we pursue its conclusions. In sum, the time is right, the audience is right, and what we need to do is get down to business.
>
> Lt. General Lawrence Skantze, Air Force Deputy Chief of Staff for Research, Development, and Acquisition to Senior Executive Conclave on Manufacturing Technology/Modernization, Dayton, Ohio, July 1980.

Whenever Congress and the public buy a tank, missile, or ship they are investing in a production technology as well as a weapon. Increasingly they are investing in automated production systems, a phenomenon that has drawn little publicity and almost no public debate, despite the fact that the more high-tech—and visible—the weapons system, the likelier it is to be manufactured in part or in whole by highly automated means. Indeed, the way these systems are produced has made the military the most significant sponsor of manufacturing automation in the country.

When touting its automation efforts, the Pentagon drapes itself in the mantle of productivity improvements and economic expansion as well as national security. But the Pentagon's rhetoric and deeds raise more questions than answers about the impacts of defense industry automation. Those impacts—on industry, on regional development, on educational institutions, on the size of the military budget, and, most importantly, on people directly and indirectly tied to the military production system—constitute the considerations of this chapter. To begin exploring the Defense Department's role in manufacturing change—and how it affects relationships between labor and management, public and private spheres, worker and manufactured goods—it is useful to glimpse the long history of automation, military-style.

DEFENSE AUTOMATION, PAST TENSE

In 1898, Frederick W. Taylor took a position with Bethlehem Steel. Taylor was the father of scientific management, a radical new technique for reorganizing shopfloor design, dividing and routinizing the individual tasks of skilled workers, and adding layers of managers and expanding their power over the production process. Bethlehem was still a relative novice to arms manufacturing, having just recently seen the financial advantage of making steel products for an engine of global growth instead of its domestic counterpart. In the mid-1980s, plant superintendent John Fritz had enthused that Bethlehem "could sell a small amount" of armor plate for warships for "hundreds of dollars per ton instead of a large amount in the shape of rails for tens of dollars per ton."[1] A few years later, the company was not only selling armor plate but attracting a congressional investigation into its peculiar price list; armor cost the Russian navy only $250 per ton while Washington way paying $600. To deflect criticism and to meet production demands created by the splendid little war against Spain, Bethlehem rushed in Frederick Taylor.

Taylor brought with him a piecework system that, he warned, would take many "details connected with the running of the machines and management entirely out of the workers' control."[2] Taylor designed his system to combat "soldiering"—workers sticking together to control their pace and output. But during the Spanish-American War, Taylor's challenge to the shopfloor status quo provoked at least as much resistance from management as from the workers. "I got into a big row with the owners of the company on that labor question," Taylor wrote. And then added:

> They did not wish men, as they said, to depopulate South Bethlehem. They owned all the houses in South Bethlehem and the company stores; when they saw we were getting rid of labor and cutting the labor force down to about one-fourth they did not want it. They came to me and said so frankly, "We don't want that done." I said, "You are going to have it, whether you want it or not, as long as I am here."[3]

The statement was vintage Taylor—abrasive and supremely self-confident—but it spoke to a more universal audience than Bethlehem's managers or the steel industry. Among others, the U.S. military was listening. Between the turn of the century and the beginning of World War I, scientific management was introduced into numerous navy yards and army arsenals by Taylor himself, by his immediate disciples, or by military officers imbued with the principles of scientific management.

At government installations, scientific management provoked strikes and a revolt by navy line officers against their Taylor-converted superiors. Again, whether they wanted de-skilling, job reductions, new forms of discipline, or not, they were going to have them as long as Taylorism was there.

Technology, Taylor-Style

Twenty-five years after Taylor's death, the values of scientific management informed wartime developments in such fields as electronics, servomechanisms, computers, and emerging theories about communications and control of information. Midwifed by World War II's unprecedented production needs and an equally unprecedented flow of government funds to the technical community, those developments contributed to the most sophisticated generation of weaponry ever created.

After the war, this new hardware and software (and theoretical outgrowths such as operations research and systems analysis) were all applied to the huge command, control, and communications challenges posed by the Pentagon's internal management and new global mission. But these technologies suggested significant changes in manufacturing as well as military capability. In many cases, new process and fabrication technologies spilled over from weapons developments. But in other instances the military expressly set out to automate its contractors' facilities.

Numerically controlled devices (N/C) were probably the best-known of these postwar automation efforts. N/C transferred repetitive machine tool operations onto a memory tape. Developed at M.I.T.'s labs, it foreshadowed the automation of batch manufacturing instead of the continuous, high-volume processes (such as petroleum and chemical refining, papermaking, or uranium enrichment) that had been the focus of most prior automation efforts.[a] The air force not only subsidized N/C research but singlehandedly created a market for N/C when it became apparent that few companies would buy the controls for commercial use.

Devoting an estimated $62 million to the effort, the Air Materiel Command ordered 105 numerically controlled machines to be placed in prime and subcontractors' factories, selected four joint ventures from the somewhat recalcitrant machine tool industry to make the equipment, and insisted upon standardization of both hardware and software. APT (Automatically Programmed Tooling language) inventor Douglas Ross called the software standardization campaign "the world's first major cooperative programming venture."[4]

Cooperative though it was, the standardization push resulted in the selection of "perhaps the most complex and expensive approach to N/C then available."[5] It mandated the use of a punched-tape format (rather than magnetic tape) that was prepared on a computer (instead of manually on a Flexo-writer), but it was difficult to read (except for a format machinists who could readily decipher it). Moreover, the benefits of this effort redounded almost entirely to the largest companies in the computer, machine tool, and control systems fields, as well as to a limited number of big aircraft firms.

a. Some observers differentiate processes like brewing and steelmaking, which work with batches of material, from "discrete products" manufacturing, which can encompass aircraft, apparel, and machine tools. In this chapter, we lump both types of manufacturing under the lable "batch" or "job shop" and note various estimates that 25 to 95 percent of all U.S. manufacturing falls into this category of lower volume, noncontinuous production, and that the percentage will grow in the future.

Smaller companies and promising alternative softwares were either foreclosed or discouraged by the combined efforts of the air force, big business, and M.I.T.[6]

The N/C and scientific management revolutions differed in several important ways, but both grew out of similar conditions—military preparations, growing trade-union movements, and periods of corporate concentration and internationalization.[7] And N/C was not only technically dramatic on its own terms, but heir also to Taylor's desire to displace human labor and to strip workers' institutional and workplace power. "The whole purpose of N/C is to remove the operator from the process," said managers at Brown and Sharpe, one of the country's largest machine tool makers. "We don't want the operator to intervene except to offset initially [i.e., make tool offset adjustments]."[8]

"Numerical control is not strictly a metalworking technique," *American Machinist* rhapsodized in 1954, "it is a philosophy of control."[9] Ultimately, N/C, like Taylorism, fell short of its proponents' fantasies of control and "depopulation." As historian David Noble's forthcoming examination of N/C shows, the goals of automaticity, de-skilling labor, and expanding management power turned out to conflict with the goals of productivity and quality.[10] But N/C's development and use etched indelible patterns for military intervention in the nation's production processes.

BUCK ROGERS WITH A LUNCHPAIL: THE SCOPE AND NATURE OF MILITARY MT PROGRAMS

Over time, an expanded air force's N/C program became a general effort to refine manufacturing technologies commonly called ManTech (MT). The army and navy launched comparable programs, and in 1975, former Texas governor William Clements, who was then Deputy Secretary of Defense, initiated a Department of Defense (DoD) effort to coordinate the three services' MT programs. A "Statement of Principles," prepared by DoD in 1980, encouraged multiservice investments, like the triservice effort, to automate electronics production, in addition to other similar endeavors. Between 1977 and 1981, the military invested approximately $640 million in MT.[11]

Today, the army manages about 600 ManTech projects worth a total of $300 million. The navy has budgeted $112 million for productivity schemes at its repair depots between FY 1981 and FY 1985. In 1981, DoD announced it would spend twice as much on the services' extended family of MT efforts over the following five years as it had in the preceding five; it then revised the projection upward in 1982 to $1.6 billion. Projected industrial preparedness expenditures of $2 billion on production facility improvement and modernization between 1982 and 1986 dovetail neatly with the MT programs. What started in the post-1945 period as N/C is now a sprawling combination of efforts to automate defense production.[12]

Measuring the significance of the Pentagon's automation programs by the number of acronyms or dollars invested tends to under- or overvalue what's actually there. It is fair to say that over the past thirty-five years, DoD's automation subsidies have: (1) involved strategic technological choices (like N/C and computer-integrated systems) that promised major structural change—both for industry and labor—for relatively little money; (2) profoundly influenced the language used to describe new technologies and the language (software) used by those technologies; (3) coordinated, standardized, and legitimated changes in various production processes as no other government agency and few oligopolistic industries have been able to; and (4) developed and promoted technologies that delivered much less than they promised. Today, seven characteristics distinguish DoD's MT investments.

1. Most investments focus on weapons production (although the army is automating such processes as the inspection of textile colors and the production of insulated combat boots). Weapons systems give DoD a context to pursue the coevolution of new components, materials (like composites), and manufacturing technologies.[13]

2. ManTech seems to manifest the American passion for push-button warfare. For example, Norman Augustine, a Martin Marietta executive who chaired the 1980 Defense Science Board study on industrial responsiveness, envisions a generation of "brilliant missiles" succeeding the current crop of "smart" weapons. "Over the years the IQ of tactical missiles has gradually been increasing," Augustine wrote in October 1982.

> Yet these missiles still rely on humans to look at the image created [on infrared and electro-optical scanning systems], recognize a target and point it out to a missile. The question arises: Why not build a missile that can by itself

> tell a tank from a rock or perhaps even from a truck? Imagine what such an autonomous missile could do. A GI could tell the missile to fly a hundred miles or more to a specific site where, say an enemy armor formation was reported. By voice the GI would instruct the missile to give first priority to tanks, second to armored personnel carriers and third to trucks . . . the missile, recognizing the voice commands, would proceed to do exactly as instructed, destroying not one but perhaps five tanks.[14]

Augustine acknowledges that "the simple matter of seeing and recognizing is proving to be an exceeding intractable undertaking [in missile design]." The new twist is that artificial intelligence technologies developed for factory automation are now as likely as battlefield automation research to make the project more tractable. "Research into this field is burgeoning," Augustine notes, "spurred by pressures from such diverse applications as X-ray interpretation and robotic factories."[15]

3. The Defense Department sponsors the automation of some continuous processes, such as the making of solid-fuel propellant for missiles. But the Pentagon directs most of its programs at batch manufacturing, the mode of production for most big-ticket weapons systems. In aerospace, for example, a Lockheed or General Dynamics will produce only ten or twelve aircraft a month, each bearing unique specifications. Batch manufacturing has particular labor requirements. Some members of the military-industrial community acknowledge that direct labor accounts for only a tiny fraction of total batch-manufacturing costs—10 percent, many observers say, in contrast to the 60 to 70 percent taken up by such indirect costs as planning, scheduling, and controlling operations and people.[16]

The world of military MT is rife with idle talk about automating that 60 to 70 percent lump of white-collar work. But DoD sells its programs as responses to purported deficiencies in the direct labor force.[17] It gets support from people like James Gray, president of the National Machine Tool Builders Association, who says that the U.S. faces "one of the greatest skill shortages in the history of the country."[18] "It is not possible to hire the experience that you lose when an employee retires," explains *Iron Age.*[19] Automation compensates for a declining workforce. However, in the parts of the metalworking business populated by Gray and *Iron Age*—sectors that form the defense industry's backbone—none of the talk about "shortages" and "decline" appears to be borne out by hard data, according to the Labor Department.[20]

4. Correspondingly, the Pentagon defends its automation programs as remedies to the prime contractors' inability and unwillingness to make investments themselves. During the 1980s, capital investment for all U.S. industry averaged 8 percent of sales, but defense contractors invested only 3 percent.[21]

Gary Denman, chief of the Air Force Materials Lab's Manufacturing Technology Division, says that contractors "clearly would not" make MT investments without taxpayers' money eliminating the risks. Subsidies "get us two to five years ahead of the game," Denman estimates. "Their [contractors'] capital decisions are not based on what they think DoD is going to want in 1986," Denman says. "It's what they *know* DoD is going to want in '83 and '84."[22]

"ICAM [Integrated Computer-Aided Manufacturing] is an air force program because no single industry or program has the resources to undertake it alone," the service trumpted in 1982. "The Air Force couldn't wait for these systems to evolve naturally."[23]

In fact, ICAM and associated efforts are air force programs because contractors know how to exploit a free lunch. "We have contractors set up just to get ICAM programs," says Lt. Gordon Mayer, and ICAM officer at the Air Force Materials Lab. "We're keeping them alive. People are automating for automation's sake in several cases. There is no good reason, there is no justification and in fact it may be detrimental. We work with parts of companies whose job it is to implement these advanced technologies. And if they can get a project from the Air Force, regardless of its real payback, they keep in business."[24]

5. Advocates boost military MT programs as efforts to build up the defense industrial base's lower tiers of subcontractors and to make military procurements more nearly resemble a free-enterprise system. But almost all DoD automation subsidies go to big business.[25] And to the defense industry's truly competitive vendors, the military's goal of technology standardization has distinctly anticompetitive features.[26]

6. ManTech program claims to greater productivity and lowered costs seem dubious. For example, the air force—known for having the military's most cavalier attitude toward MT payback—says that its TECHMOD project at General Dynamics will return several times over the government investment of $25 million. But James Ashton, the General Dynamics Vice-President of Production who oversaw automation at air force plant #4, demurs. In remarks to the National

Academy of Engineering, Ashton acknowledged the "extensive national publicity" surrounding "the most pictured robot that has ever been made," as well as the positive effects of introducing photogrammetry, a new CNC-DNC system that linked together "some major new numerically controlled machines," and new ordering, inventory, and data control systems. "However," Ashton continued, "basically not one of these [Techmod-inspired] things, or even the collection of these improvements in a narrow sense, could possibly explain" productivity changes at the plant. Instead, Ashton, argued, whatever improvements occurred were due largely not to new technologies but to a shift away from an "autocratic, short-term results, by the numbers" management style.[27]

Even more skeptical is retired General Henry Miley who directs the American Defense Preparedness Association, a leading contractor trade group. "Our association has been involved with many ManTech [manufacturing technology] reviews and ManTech symposia," Miley told a House Armed Services Committee panel in 1980. "What concerns me is that when I get up and raise my Yankee voice and say, can I go out to some factory and put my hands on an item that is being produced more cheaply now than it was give years ago because of the ManTech, I get kind of a confused answer."

Actually, as General Miley revealed minutes later, the answer he got was fairly straightforward. "When I asked the bottom line question, is the missile now cheaper than it was two years ago [because of ManTech], the answer was, well, no."[28]

7. Sound or not, MT programs put DoD on the cutting edge of a changing international economic order. In the past, N/C and other projects, like laser welding and diffusion bonding, were rationalized by military production needs and couched in terms of shopfloor efficiencies. But today, with America's status in the international economy a very live issue, DoD increasingly promotes its MT efforts as a boost to the country's global market-share and reputation for industrial leadership. Military MT is to drive *all* U.S. industry, not just predominantly defense sectors, toward a more competitive posture.

With industrial supremacy a prime military objective, the opponent who galvanizes manufacturing change is not the USSR but America's allies and trading partners, particularly Japan and Western Europe. "The Air Force saw that flexible manufacturing technology was springing up in Japan and Europe and wanted to find out why the technology was not being widely applied in the U.S.," an Air

Force ManTech official told *Iron Age.* "So we asked a number of key officials in the industry to meet with us to discuss the matter and we learned that U.S. aircraft and aerospace companies were having a hard time getting a handle on the concept. That's really when we decided to go ahead with the project."[29]

MILITARY MT AND REGIONAL DEVELOPMENT: THE SOUTH

The political environment that the Department of Defense encounters in America's southern states mirrors the global struggle to protect and attract investment dollars and market-shares. Like their counterparts in every other part of the United States, southern elites can barely contain their belief in the curative powers of science- and technology-based industry.[30] Thus far, DoD hasn't helped very much. The South has hosted even less Pentagon-sponsored research and development than it has high technology weapons production.

Eight southern states[b] (which contain 15 percent of the U.S. population) received only 5.4 percent of the total $10.3 billion in 1981 prime contracts (over $10,000) that DoD awarded for research, development, testing, and evaluation (RDT&E). Moreover, nearly all the RDT&E contracts in Virginia were awarded to "beltway bandits," a bizarre tribe of Washington-area firms with its own rituals and language that belongs to no region. Excluding them, the eight states received roughly 2 percent of the nation's RDT&E prime contracts—four and one-half times less than Massachusetts. It would be fair to say that DoD has not exactly perceived the upper South as the hub of the knowledge industry.

MT Goes to School

Increasingly, the RDT&E money—and, to some extent, the production contracts—that DoD *does* put into the region are earmarked for MT. And more and more the place where the Pentagon's automation agenda and state and regional development strategies come together is the university.

b. Alabama, Georgia, Kentucky, North Carolina, South Carolina, Tennessee, Virginia, and West Virginia.

The Defense Department's effort to spread the MT gospel throughout the academic land is intensifying at exactly the time when private and state universities are being used as industrial recruitment bait and industrial seedbeds as never before. In 1981, the Air Force Systems Command kicked off a $3 million Manufacturing Science program, designed to complement a similar effort by its Office of Scientific Research in "stimulating greater involvement in manufacturing in academia."[31] According to air force ManTech chief Gary Denman, the new program will serve as a vehicle for tapping nonengineering disciplines, for example, "the physicist who may have an idea for a device but no independent engineering background."[32]

At the same time, DoD is pushing to consolidate manufacturing disciplines within the academy. Currently only three universities in the United States offer graduate degrees in manufacturing engineering. "Manufacturing engineering has been the application of all kinds of engineering to manufacturing," says Dr. Arthur Thomson of Cleveland State University.[33] But these patterns are changing, due in no small part to DoD's willingness to spend big money for lab equipment and to recast the connections between its commands, production contractors, and universities.

One symptom of the change is the new manufacturing systems engineering major Dr. Jay Black is setting up at the University of Alabama, Huntsville (UAH). Black has set his sights on the $150 million that DoD will dish out between 1983 and 1987 to equip projects just like his.[34] So, he says, are most of his colleagues. But well before the five-year equipment grant program was announced, UAH and the Army Missile Command (MICOM), located at nearby Redstone Arsenal, had begun reshaping their relationship with a fresh emphasis on manufacturing.

Black, who served as a consultant to the triservice Electronics Computer-Aided Manufacturing project, says that historically DoD's MT programs "had a charter which says 'we really can't go invest in research.' So if you were able to extract some DoD dollars you did it surreptitiously. You said that you were going to do this for DoD and by gosh, you did it but you also did some research on the side." Now things have changed. "There is tremendous awareness on the part of the folks out on the arsenal that [MT] is important stuff. We would become a grantee specifically to look at how we could improve the productivity of a particular form of manufacturing, say electronics."[35]

Like everyone else, says Black, the state of Alabama is pushing high-tech growth. "Alabama is about ready to make a jump but we don't know which way we're going to go yet. Of course, robotics is the magic word. Most people don't know what the hell they'd do with one if you gave 'em one, wouldn't know how to put it in if they had to but, by God, they're going to get one if they possibly can."[36] Presumably, electronics automation research at state-supported UAH would benefit MICOM's out-of-state contractors more than it would Alabama's under developed electronics industry. However, Alabama has made no overt effort to graft its industrial development agenda onto Jay Black's new bailiwick. "Beats the hell out of me," Black says of the state's designs. "I don't know the answer. I don't know whether they're asking the question or not."[37]

Alabama's neighbor, Georgia, offers a similar example of the collision of military high-tech spending, state and university development agendas, and the public interest—a drama that has been played out at Georgia Tech, the Pentagon's major research and development outlet in the Southeast.

Georgia Tech: Military versus State Development

Tech and the Lure of DoD Dollars. Georgia's legislature created Georgia Tech's Engineering Experiment Station (EES)[c] in 1919 as an industrial counterpart to the state's agricultural experiment and experiment and extension apparatus. During World War II, EES came into its own, working on various aeronautical projects, including helicopter design and testing; expediting the use of southeastern minerals (by developing ways to substitute olivine for magnesium and use filler clays in synthetic rubber); and researching the materials characteristics of quartz crystals used in the Signal Corps' radio and radar equipment.[38]

Early in the postwar period. Tech's research assumed a schizoid character. On the one hand, the EES did its development duty by performing R&D in such areas as the mechanization of peanut processing. But Tech also tried to prolong and expand the World War II research contracting boom. Budding industries, like aerospace, of-

c. Georgia Tech's Research Institute, the formal recipient of most DoD funds at the university, acts as the EES' contractual and patenting agency.

fered one vehicle. Modernizing the school's wind tunnel, for example, enabled the EES' Mechanical Sciences Division to double the contracts it took in between 1956 and 1957. Lockheed-Georgia helped pay for the tunnel's modernization, and serving Lockheed blended the institution's state and contract development missions perfectly.[39] However, the major thrust of EES' contract research, from the war onward, was electronics; its principal customer, the military. While the Air Materiel Command was calling upon M.I.T. to help develop N/C, it was lavishing awards on Georgia Tech that eventually made the school preeminent in electronic warfare research.

The quandary created by Tech's dual mandate—economic expansion for the state and electronics research for the military—was clear to outsiders as well as the people who ran the university. In 1963, the Arthur D. Little Company prepared a study for the school's Alumni Association that attempted to resolve what it called the "internal friction" resulting from the university's competing research missions. Little buttered up the alums by portraying the "dilemma" as idealized ivory towerism in conflict with Tech's industrial duties. But in its report, Little sought to resolve the real dilemma, which stemmed from Tech's growing concentration on military electronics and the persisting lack, in Georgia, of industrial investments or jobs created by that research.

The EES "might expect to sell" its research services to "the growth and defense industries whose operations are based on changing technology . . . [particularly] if the research departments of these industries were also located in Georgia," Little contended. But Georgia had neither "head offices" nor research departments, and Atlanta was growing as a center of finance, retailing, service industries, and transportation, not manufacturing. "Until Georgia's [undiversified] manufacturing base is strengthened," Little concluded, "the EES cannot expect a substantial increase in contracts from the private sector of the state." Little advised Tech to keep pursuing federal contracts but also recommended that "efforts to sell commercial contract services be extended to the larger market beyond the region."[40]

The university acted on both recommendations. The EES' contract work for out-of-region corporations soared, along with DoD awards. But Tech also fulfilled the circularity of Little's prescription. Georgia lacked high-tech manufacturing but Tech's DoD-nurtured expertise lay in that area. Marketing its expertise outside the state and region would build up the EES. But doing so also reinforced

Georgia's reliance upon recruiting runaway, low paying, labor-intensive manufacturing industries and building up Atlanta as a service center. It wouldn't help establish a more diversified (and better paying) manufacturing base. The Little report proposed a way out of this circle—the creation of an industrial research park for indigenous science- and technology-based industry. But nothing came of the recommendation.

Tech spun off an occasional company like Scientific Atlanta but relied almost entirely upon random industrial recruitment to build the state's base of high-tech industry. It was a gamble that largely failed. "When people think of high technology they don't immediately think of Atlanta," says Wayne Hodges, associated director of Georgia's Advanced Technology Development Center (ATDC). "Immediately? I guess you could amend that to say people don't think of Atlanta when they think of high tech."[41] For years, in fact, Georgia has had to point lamely to the EES itself as "one of the largest high technology employers in the state and among the largest in the South." EES employs 700 people.[42]

Tech and the MT Bandwagon. Hodges's ATDC is part of Georgia's larger campaign to use Tech's military dollars not only to build the EES but also to develop a manufacturing base featuring new process and product technologies. In 1975, Georgia's General Assembly created the country's first state institute to aid corporate productivity. Housed at the school, the Productivity Center has concentrated on robotics and CAD/CAM and set up a Materials Handling Research Center with support from the National Science Foundation and two dozen corporations.[43] The same type of T-3 robot that established TECHMOD at General Dynamics has been borrowed from Cincinnati Milacron and installed at Tech's School of Mechanical Engineering to "speed up the introduction of this sort of equipment into Georgia industry," according to Rudy Yobs, the Productivity Center's director.[44]

The ATDC, established shortly after the Productivity Center, helps entrepreneurs develop, finance, and commercialize new high-tech products. It is laying plans to create a Robotic/Automated Manufacturing "Center of Excellence." ATDC claims that "high technology industries are among the first to employ these new methods," that "Georgia is ideally placed for growth in this area," and that "if the products [robots and CAD/CAM equipment] are available from

Georgia firms, they will be used."[45] Lockheed and the air force ICAM project have sponsored Tech research in sheetmetal-working technologies. But to date, comparatively little of Tech's DoD funding has been for manufacturing technologies—"probably because we're coming from a relatively small base," says Yobs. "It takes a certain amount of time to get up to speed and make your beachhead."[46]

One potential beachhead is automating electronic production—a major goal of both industry and the military. Tech's EES has joined the electronics automation project committee of CAM–I, and the EES' 1982 annual report anticipates that the CAM–I connection "may lead to projects for EES."[47] DoD's ECAM, the other major, coordinated effort to automate electronics production, just concluded a $2 million planning phase identifying products like printed circuit boards, cables and harnesses, hybrid microelectronics, and integrated circuits as targets for automated production.[48]

Tech's interests, however, range far beyond research contracts. The EES is finally moving to establish an electronics industry—an automated electronics industry—in Georgia. In 1982, Tech and the ATDC produced a preliminary blueprint for just such an enterprise.[49]

"I think the automation of electronics production is going to become extremely important," Rudy Yobs says. "You'll begin to see more merging of our efforts as manufacturing engineers with some of the electronics types, I'm sure. I think it permits us to capitalize on [our electronics research] strength and all the manpower we have here." Yobs also says that "this is a type of industry in which a lot of new jobs may lie for Georgia."[50] Other states, however, have gotten an earlier start on electronics manufacturing will provide fierce competition, and the number of jobs to be had anywhere is unclear. Automating the production of electronics components, after all, will be a decisive step toward automating the production of most labor-displacing factory technologies, an integral part of the spiral that already has enabled a Japanese manufacturer to build robots with robot labor.

What is clear in the short term is that Tech—one of the state's most important economic instruments—sees DoD's emphasis on MT brokering both new commitments within the university and broad ripples throughout the state's manufacturing economy. "The concern is there at DoD," says Yobs, "and they are spending substantial amounts of money. We hope that we will be able to secure support there for [MT] research that will sort of serve a double purpose—

that will help build our own capability which we can then apply to some civilian problems here."[51]

Development, Jobs, and Productivity

Georgia Tech is hardly alone in planning short-term and approaching the long run on a wing and a prayer; pursuing factories of the future is considerably easier than projecting their impact on regional development. Many, like air force ManTech director Gary Denman, insist that automation will make the U.S. labor market more competitive with Third World exporting countries where work comes cheap. "When things are not labor-intensive [i.e., when they *are* automated], they have not shifted out of the country," Denman argues.[52] And some members of the engineering community claim that the cost and extensiveness of new MT investments will act as a brake on corporate runaways inside the United States. On the other hand, *Fortune* reports that "the more automated production becomes the easier it can be shifted to countries where labor is cheap. Several labor specialists suggest that if $12-an-hour labor in the U.S. and $4-an-hour in, say, Brazil can do the same work with the help of robots, auto production will inevitably shift out of the U.S."[53]

The Defense Department, as the great subsidizer and standardizer of automated production technologies, plays a crucial role in redefining the industrial location process. Acting like a national product-transitioning center, DoD speeds various manufacturing technologies from the shelf, the drawing board, or experimental usage into factories. Immune to the obstacles stymying new capital investment, DoD diminishes both the time involved in routinizing new process technologies and the importance of locating factors, such as proximity to great universities, markets, or sister companies and industries.

But DoD also subordinates questions about automation's long-term, macroeconomic impacts to the fulfillment of short-term corporate and military agendas. DoD's subsidies may ultimately invite transnational companies to ship all manner of production out of the United States to Brazil and other locations. But thus far, that possibility has not even entered the discussion about security objectives and defense industrial capacity. Nor have more immediate questions about labor market impacts—the most logical questions to be asking at this juncture—been part of the discussion.

Air force officials in charge of the ICAM program say they simply haven't looked into the extent or effects of displacement that could result from aerospace automation. Indeed, a 1980 Air Force Systems Command report, *Manufacturing Technology Investment Strategies*, estimates a 44 percent reduction in "people" as one of its "CAM center benefits."[54] The number of jobs generated by *all* types of high tech, including the manufacture of production equipment, has been questioned by even such previously unabashed enthusiasts as *Business Week*—among many bubbles the magazine's March 28, 1983, issue pops being Georgia's rosy high-tech job forecast.[55]

Ultimately, the questions about where the jobs are and how many automation provides and how many it takes away come full circle to questions about what kinds of jobs, who decides, and whether the new way really is more productive.

The impact of new technologies on job content is seldom raised at Georgia Tech or air force plants. But people with power to make technological decisions have expressed themselves quite clearly on the subject. When the air force surveyed contractors during ICAM's formative years, for example, industry executives "considered management control as having the greatest payoff potential in CAM."[56] The National Machine Tool Builders Association, harbinger of the "skills shortage," takes a corollary position. In a report prepared for the air force, NMTBA trained director, John Mandl, proposed to cut machinist apprenticeship time by as much as half, "upgrade the content in a narrower scope . . . [and] *reduce the skill levels* required to operate or maintain certain machine tools" [emphasis added].[57]

Despite NMTBA's intention to de-skill, and management's perchant to imprint social control in new technologies, the rubric of 1980s defense industry automation litanizes job "enrichment," the replacement of "tedious" handwork with what one academician calls "the electronic manipulation of symbols,"[58] and—in some quarters—limited worker contributions to decisionmaking.

"PANDA (Profile and Drill Automation) is the technology of the 1980s," enthuses Lockheed-Georgia's director of manufacturing technology, Joseph Tulkoff. "The old way—back-breaking, archiac—was the technology of the 1930s, pushing all day with hand labor. In the new way, the machine is doing the back-breaking work and the worker is doing the thinking. This makes the job more interesting, more challenging."[59] Richard Wysk, of Virginia Polytechnic Institute & State University (VPI) says, "Everybody's grade in a factory is going to be upgraded. You used to be a machinist, you will now become a parts programmer. You will generate the tape that is required for it."[60]

ICAM's Gordon Mayer goes a step further, suggesting that production-worker involvement at the design stage of a new technology is "not only feasible, it's essential. We've not involved them in the past and they've laughed because they knew we were screwing up the whole time. They knew something that wasn't written down—most things aren't. They had the corporate memory."[61]

Despite these lofty and pragmatic sentiments, the recognition that workers have uniquely useful things in their heads is *not* guiding the design, introduction, and usage of defense industry automation. To the contrary, new military production technologies are frequently being implemented much the same way N/C was in the 1950s—in spite of or as an attack on production workers' intelligence, skills, and unions.[62]

That attack, and the way military values and goals color it, are revealed in places as different as VPI and Lockheed-Georgia. Virginia Tech recently participated in the air force's sole attempt to delve into the human side of production technology issues in ICAM—a Honeywell-led "human factors" study of three automated systems. Lockheed-Georgia has been a repository of new manufacturing technologies since the 1950s.

By definition, the study of human factors elevates the importance of human needs in technology design and application. But the goals are distinctly limited. "We deal with interfaces, controls, displays, physical environment, workspace design," says one human-factors engineer. Human-factors engineering dates to World War II, when it was used to adapt such things a gunnery control systems to enable artillerists to keep up with the new speed and performance capabilities of their weapons and the objects they were built to shoot down. Now, human-factors engineers take aim at such machine or workplace problems as cathode ray tube (CRT) glare and the harshly hyperbolic error messages ("ABORT!!!!") incorporated in many computer systems.

Between 1979 and 1981, ICAM's Human Factors project looked at Hughes Aircraft's ICAM Decision Support Systems (IDSS), a process-planning aid; at a General Utility System (GUS) for simplifying human-computer interactions; and at the TECHMOD robotic cell at General Dyanmics' F-16 plant in Texas.

The study briefly noted and carefully danced around vital union-management issues at General Dynamics—just as it had danced around the prospect of an equal three-way partnership in executing the study. Gordon Mayer, director of the Human Factors project, says, "We involved [the International Association of Machinists] not as a union per se but as the workers of that union. We talked with a lot of people. But to specifically answer your question, no, I don't think the union was involved officially."[63]

ICAM's Human Factors group interpreted their observation that companies "added layers of management" to handle automation to mean that automation *necessitates* additional management. The study noted the absence of literature about unsuccessful automation experiments but confessed that "our interviews resulted in our continued inability to tell what the critical difference was" in automation successes and failures. And the $225,000 project report ends with such breath-taking conclusions as "human factors occur throughout computer-integrated manufacturing" and a call for "careful information and training programs."[64]

Lockheed-Georgia occupies the world's largest aerospace plant—equal to the size of seventy-five football fields with a sixty-two-acre farm left over—and the air force–subsidized N/C equipment it received in the 1950s has been supplemented with various ICAM and TECHMOD investments, including a sheetmetal cell, in the '70s and '80s. The plant is a real-world incarnation of VPI's "table top." Larry Strugill (pseudonym) is a veteran Lockheed machinist whose milling machine has just been outfitted with a terminal tied to a central computer in Lockheed's basement. "They've gone from the tape (N/C) machines to the computer numerical control machines and now they're moving towards all the machines being run on the central computer," Sturgill says. "You have a computer terminal at the machine and you just punch in whatever job you're doing. The computer downstairs will load and disc with the instructions for that job onto your machine." Coupled with the central computer are layers of supervision that turn the Lockheed tableaux into "people watching people watching people watching people watching machines."

Sturgill's machinist lodge is one of the biggest union locals anywhere in the country. Partially because of the union's power and partially because of military contracting's dynamics, the wave of technological change at Marietta hasn't yet caused massive displacement, despite what Sturgill calls "a certain tendency to think that anything you can do that will eliminate a worker is good." Like Bethlehem in 1898, Lockheed has other investments to protect; today, it's a grass-roots lobbying force instead of a company store. "As long as they get money from the Air Force they're not that concerned about trimming the payroll," Sturgill explains. "Maybe to a certain extent there are political advantages to keeping a lot of people on the payroll even if they're not working that hard. They always try to mobilize us when they go after these different contracts. It's certainly not to their advantage to conspicuously lay people off when they're right at the height of going after a contract."[65]

PROGRAMMING MACHINES: PROBLEMS WITH THE MANTECH FUTURE

The following are excerpts from the interview with "Larry Sturgill" of Lockheed-Georgia, a skilled machinist who has had some experience with manufacturing technologies—in this case, tooling machines operated by a computer-programmed tape.

> The idea behind the tapes machines is to have absolute uniformity. Particularly they try to minimize any kind of interference with the program by the operator. But if the program isn't made right in the first place, they've made it difficult for the operator to interrupt the tape and go back to find out where he is. If the program doesn't work perfectly, they've made it inordinately complicated to change anything.
>
> Most of the thinking is supposed to be taken care of in programming. The operator is just supposed to clamp the part in the machine and press a button and start it up. The philosophy behind it is that the operator's the least smart person. So if you let the operator go messing around with anything, he's gonna screw the part up.
>
> In practice, the problem when you have the tape machine is that you end up with such a bureaucracy. You have one group of people making the fixture that the machine is going to be on, tooling, then you have somebody who is programming the part. . . . The engineers who design the fixtures have probably never worked in machine shops. The programmers are mostly people who have a lot of computer experience but not people who've ever worked in a machine shop. . . . And the tooling and programming and production departments don't work together well.
>
> If you change just one little move or a position, that might affect something else in the tape; so there's a whole procedure you'd have to go through. You have to get an inspection, you have to go over the part again. So usually they find some way of making an adjustment by

manually moving the machine or changing the fixture so they can get around changing the program.

At any rate, the operator usually ends up the one who is responsible for trying to figure out how to correct all the mistakes that were made in tooling and programming and make the part work out right. . . . A lot of times they end up making kind of a crude part on a machine; then somebody outside the shop has to sand the part or file it or do whatever is necessary, or just put it on a conventional milling machine to get the dimensions that it was supposed to have in the first place.

It's hard to say whether the program is bad or not. I think it's more probable that the principle behind programming is somewhat erroneous. If you were a time study engineer and went into a machine shop the two things you would observe the machinist doing would be making calculations—he sits down, looks at the blueprint, does the calculations—then putting the part in the machine. He's mostly positioning it, moving something from part A to part B, or moving a cutting tool. If you look at it superficially, that's what's involved in machining and you should be able to duplicate it. . . .

The problem is that there are a lot of subtle things in machining. The skill of the craftsman is not apparent. If you watch somebody make a piece of pottery, the pot's a simple curve, you just have a wheel that's cranked with a foot pedal. And if you hired a team of engineers and programmers there's no doubt that they could probably, with enough money, make some kind of a machine that would make a pot. But it would be an inordinately complicated machine. And the pot would probably not be a very good pot.

In machining, there are even a lot of subtle things in drilling a hole. All you can tell a machine is that you start to drill at this point, you go in so deep and you come back. But you can't tell a machine that if there's a hard spot in the metal it should push through, or if it starts getting overheated it has to start coming back out.

It's not so much that the programs are bad. It's just that its impossible for somebody sitting in an office somewhere to try to write up a set of instructions and binary codes for a machine to do something like that.

The issues raised by defense-industry automation are not new—How many jobs? Where will they be? What will they be like? Are the new technologies really more productive (in the narrow sense of making more things faster and cheapter and in the fuller sense of making producers more resourceful, creative, and happy)? If so, productivity for whom? But although the military seeks credit for leading the reindustrialization of America, it has essentially refused to reckon with the significance of the above questions for the defense industry and society at large.

While no voices inside the military-production establishment have begun this reckoning, the silence outside the military is just as deafening. The General Accounting Office (GAO) appears to be virtually the only part of the federal government remotely interested in DoD's automation programs. In its public pronouncements thus far, GAO seems principally concerned with better coordination with DoD's MT efforts and other federal programs to improve automation systems. A substantive review of the impacts of DoD automation efforts seems nowhere near the top of GAO.s priority list. The congressional Office of Technology Assessment's study of automation poses many of the same questions, but fails to analyze DoD's impact.[66]

The ultimate issue raised by the Defense Department's role in factory automation is exactly the same one suggested by its part in strip-mining for coal or titanium, making body bags, or nuclear weapons: Does the Pentagon really make us more secure? If so, at what cost? And who pays the price?

APPENDIX

A MISCELLANY OF MANUFACTURING TECHNOLOGIES, MILITARY MT PROGRAMS, AND MT ORGANIZATIONS

Despite interservice coordination, the air force still stands out as the most aggressive subsidizer of military production technologies. It spends about $100 million a year on such programs. In 1977, after four years of study and promotion in industry circles, the air force unveiled its most significant follow-up to numerically controlled devices (N/C), an Integrated Computer-Aided Manufacturing, or ICAM, program.

ICAM sought to combine the latest advances in N/C, robotics, and computers, working toward the day when "computers and machines can be made to work together with little human intervention" all the way down an assembly line. With flexible, programmable machining already developed, ICAM focuses on ancillary areas of production like inventory, conveyance, and the organization and dispersal of information. Between 1977 and 1982, the ICAM program awarded more than sixty-five contracts involving some fifty U.S. companies. The Air Force Systems Command (AFSC) expects to spend $106 million on ICAM through 1985.

Currently the AFSC has twenty ICAM projects underway, including a "Factory of the Future." Led by the Vought Corporation, a team of ten contractors, including North Carolina's Research Triangle Institute, is trying to develop an entire plant that utilizes only cutting-edge production technologies. AFSC says it wants to see industry-wide implementation of the model factory's achievements between 1985 and 1990.

Along with Computer-Aided Manufacturing International, Inc. (CAM–I), a private Texas-based consortium of high-tech companies, ICAM has been acknowledged as the most prominent developer of computer-assisted design and manufacturing in the United States. The air force has never been bashful about acknowledging its own accomplishments. "ICAM took the lead in the DoD in learning how and where to use robots in manufacturing aircraft," says an AF press release. In late 1978, the ICAM program put the first "true" robot to work drilling holes in components for the F–16 aircraft at AF Plant #4, General Dynamics Corporation, Forth Worth. The robot—a Cincinnati Milacron T–3—reportedly reduced drilling and routing labor by a factor of three to one. And General Dynamics' Fort Worth plant became the stage for an ICAM spinoff called TECHMOD, as in Technology Modernization, devoted to refining specific pieces of a computer-integrated system (the "Bottom-up" approach) rather than the overall system (the "Top-Down" strategy). TECHMOD plowed $25 million of taxpayers' money into GD's F–16 production line, buying such innovations as photogrammetric inspection of assembly tools. TECHMOD is designed to leverage MT investments by the contractor, though there is no fixed rule of thumb for apportioning the burden of public–private investment.

Hard on the heels of the F–16 project, AFSC let TECHMOD contracts with Westinghouse and Lockheed. The former's Defense Electronics Systems Center in Baltimore got $7 million to automate production of AWACS radar, F–16 radar, and electronic countermeasure pods. Lockheed got $4.5 million to automate rewinging of the C–5A cargo plane. Through fiscal 1982, the air force had invested over $67 million in TECHMOD at fifteen contractors' facilities.

ICAM has sparked a variety of other computer-integrated programs like the navy's Shipbuilding Technology Program (STP), directed by the Naval Seas System Command the the triservice Electronic Computer-Aided Manufacturing (ECAM) project, managed by the army. Programs like ECAM supposedly represent new strikes in interservice cooperation (no more olive-drab robots and blue robots). The organization that has done the most to broker corporate-military connections in this field is the Manufacturing Technology Advisory Group or MTAG. MTAG hosts frequent meetings and symposia on automation and sponsors standing committees on CAD/CAM, Electronics, Metals, Munitions, Nonmetals, and Testing and Inspection. The academic community gets into the act through something called the College CAD/CAM Consortium, which has been involved with ECAM, among other projects, and seeks to introduce CAD/CAM into higher education curricula.

Flexible Manufacturing Systems, or FMS, is the name given to one family of computer-integrated technologies that is specifically geared to modest-volume metalworking *and* ushering smaller companies into the second industrial revolution. "Everybody and his brother believes that flexible manufacturing is the only way to fly," an official of the air force's FMS development program told *Iron Age*, "even though there isn't a single FMS in the United States that operates the way it was intended to."[67] Consisting of N/C machines, systems to move parts between machines and mechanisms to coordinate the machine tools and the conveyance systems, FMS's are supposed to accommodate readily changes in parts and part mixes caused by design, engineering, and demand alternations. The first air force FMS development contracts were let in fall 1981 to industry-university consortia; demonstration systems are expected by early 1986. The army is also boosting FMS. Draper Labs of Cambridge, Mass., has a three-year contract to develop a "broad-based flexible manufacturing strategy" for the Army Tank Command at General Electric's Ordnance System Plant and the Rock Island Arsenal and to produce an "FMS Buyers' Guide."

"Group Technology" is an organizational principle that, by some definitions, encompasses FMS. One physical (as opposed to administrative) manifestation of group technology is the arrangement of production equipment in "cells" defined by the shape and manufacturing requirements of different parts and designed to maximize machine utilization and expedite parts and materials flow. Often included in the rubric of group technology is "Computer-Aided Process Planning," an information-handling system that is supposed to speed up and make more reliable the work of manufacturing engineers who lay out production runs. Computer-aided process planning comes in two speeds—the variant and generative approaches. Generative planning is designed to be toally automated, summoning up the sequence of production operations and all the process parameters without sifting through prior plans. No such system exists yet, despite the best efforts of several aerospace firms.

NOTES TO CHAPTER 9

1. Daniel Nelson, *Frederick W. Taylor and the Rise of Scientific Management* (Madison: University of Wisconsin Press, 1980), pp. 77.
2. Ibid., p. 81.
3. Ibid., p. 97.
4. This summary is based upon David Noble's forthcoming *Forces of Production* (Knopf, 1984), which examines the development of numerical controls in extensive and skillful detail.
5. Ibid.
6. Ibid.
7. As the most astute historians of military-industrial partnerships have observed, World War II was hardly the halcyon time for organized labor as has popularly been thought. Indeed, a persuasive case has been made for union power diminishing during the war in relation to the combined muscle of an unprecedentedly large military (with a structured anti-union bias) and a mobilized, recapitalized corporate community. For example, see "Mobilizing the World War II Economy: Labor and the Industrial-Military Alliance," in Paul A.C. Koistinen's *The Military-Industrial Complex* (New York: Praeger, 1979). In the immediate aftermath of World War II, American employers responded to labor's newly assertive spirit with a spasm of red-baiting, claims of rediscovered "management prerogatives," and a new wave of hope vested in the perennial vision of workerless factories. Luckily for management, wartime inventions seemed to suggest that machines that would take the place of men and women were just around the corner. For example, see E.W. Leaver and J.J. Brown, "Machines Without Men," *Fortune* (November 1946).

   One area of contrast between the scientific management and numerical controls revolutions was education, a field where Frederick Taylor met with less than complete success. Taylor insisted on the quaint notion of preparing young men for the engineering and managerial professions with an educational regimen that much resembled his own background. He proposed

   > that college students be obliged, say at the end of their first year, to work in competition with men working for their living for a period of six months. The college boys to keep the same hours and have the same work that is given to ordinary workmen. I believe this will tend, by giving them a foretaste of their life's work, to

give them a more earnest purpose in their college course and also to appreciate the necessity of rigid discipline (Nelson, *Frederick W. Taylor*, p. 187).

In response, Harvard's president Charles W. Eliot sniffed that universities attempted to turn out "men having personal initiative and the faculty of independent thought" rather than "men of automatic actions like soldiers, sailors and factory operatives" Ibid.

By contrast, N/C came along at a time when the interests of the academic and corporate communities were admittedly more mutual. By the time of N/C's advent, the faculty of independent thought in Cambridge was in such a state that engineers experimenting with numerically controlled equipment at M.I.T.'s Servomechanisms Lab preferred having law students run it instead of skilled machinists. "Machinists would not trust the numbers," one N/C developed recalled, "they would tinker with the thing. The law students, on the other hand, would lave it alone and follow instructions" (Noble, *Forces of Production*).

8. Noble, *Forces of Production.*
9. Ibid.
10. While some analysts cite the relatively low adoption rate of N/C as proof of the perfidy or stupidity of American industry's capital investment patterns, it is more likely that N/C is attached to so few machine tools because it is something less than the greatest thing since sliced bread. *Fortune* claims that "fewer than four percent of machine tools installed in the U.S. are numerically controlled." See Gene Bylinsky, "The Race to the Automated Factory," *Fortune* (February 21, 1983).
11. *Aerospace Factory of the Future*, special issue of *Aviation Week & Space Technology* (August 2, 1982).
12. Various articles in *Aerospace Factory of the Future*, a special issue of *Aviation Week & Space Technology* (August 2, 1982): Clarence Robinson, "Pentagon Planning Industrial Surge Capability," *Aviation Week & Space Technology* (March 1, 1982); and Department of Defense "Statement of Principles for DoD Manufacturing Technology Program," signed by Arden L. Bement, Jr. and Dale W. Church, Deputy Undersecretaries for Research & Engineering, March 14, 1980. The army's MT efforts are summarized in "U.S. Army Manufacturing Methods and Technology Program," U.S. Army Materiel Development & Readiness Command (DARCOM), 1982. The air force's MT efforts are summarized in "Air Force Manufacturing Technology Program," Air Force Systems Command, 1982.
13. One example of this coevolution can be observed in the composite materials that substitute for various metals (see Hercules, Inc., profile in Chapter 2) and form the basis for stealth technology and performance efficiencies. Laminating graphite layers to make aircraft skins has been a painstaking, time- and skill-consuming task. But under the air force's Composite Assembly, Production, and Integration (CASPIN) program, time studies

measured on 84 percent reduction in recurring labor hours when Northrop laid up fifty-four F-18 graphite rudder skins the new, automated way. William B. Scott, "New Methods Create Composite Parts," *Aviation Week & Space Technology* (August 2, 1982).

14. Norman R. Augustine, "Brilliant Missiles on the Horizon," *IEEE Spectrum* (October 1982). Despite the historical turn from armed soldiers to manned weapons, our standing armies are still huge. In the United States DoD counts two million officers and enlisted personnel in uniform, just about as many as in 1958—a time of comparable cold war tensions but far more primitive weaponry.
15. Ibid.
16. See, for example, the "ICAM Program Prospectus," Air Force Materials Lab, Wright-Patterson AFB (September 1979), which labels the "traditional" goal of automation—the reduction of direct labor—"an ever-vanishing target." "Touch labor is only ten percent" of manufacturing cost, John Blanchard, Assistant Deputy for Materiel Development at CARCOM told a Manufacturing Technology Advisory Group meeting in Phoenix, October 17, 1982.
17. Edward Kolcum, "C-5A Wing Modification Using Advanced Methods," *Aviation Week & Space Technology* (August 2, 1982). "We're going in the direction of eliminating the manufacturing engineer," says Ralph E. Pattsfall, GE's chief manufacturing engineer for aircraft engines. Lockheed's Generative Planning Process (GENPLAN), according to E.D. Ledbetter, manager of computer-aided manufacturing systems at Marietta, "takes the place of sifting through 80,000-90,000 blueprints of detailed parts. The task—to retrieve, code, and plan a part—can be done in an hour or less when before it could take 12-16 hours. There are 7,000 operation sheets of assembly work instruction for the C-5 wing. We can now call them up in seconds."

    Others think that the brave new world of process planning is hype. ICAM's Gordon Mayer studied Hughes' air force-subsidized automatic process planning aid, the ICAM Decision Support System (IDSS) for two years. He labels IDSS "a waste of money." Mayer says, "Process planners are fairly smart individuals, I've found. In effect they've got a lot stored in their head that's not in any book. Personally, I feel when there's talk of Generative Process Planning they laugh. Because it's not going to be in their lifetime—I firmly believe it won't. A truly generative process planning system does exist but it takes a long time and it's for a very very narrow scope of parts. And it's a test basis." (Mayer, author interview, October 1982.)
18. "Why the U.S. Can't Rearm Fast," *Business Week* (February 4, 1980).
19. Alice M. Greene, "Lockheed-Georgia Advances Technology," *Iron Age* (September 7, 1981).

20. Neal M. Rosenthal, "Machinist Shortage: A Look at the Data," *Occupational Outlook Quarterly* (Fall 1982).
21. Brigadier General Joseph H. Connolly, Air Force Director of Contracting and Acquisition Policy, "What is the USAF Doing About Productivity?" ICAM Industry Days, September 1980, St. Louis, MO.
22. Gary Denman, author interview, October 1982.
23. "What is ICAM?" Fact Sheet from Aeronautical Systems Division, Office of Public Affairs, Wright-Patterson AFB (January 1982).
24. Aerospace companies that sell to commercial markets—which are far less fickle than their military counterparts often appear and therefore far more logical magnets for capital investment—haven't seen fit to take the automation plunge on their own. "Manufacturers such as McDonnell Douglas and Lockheed are acquiring automated composite cutters, material handling equipment and task-flexible robots for applications in military-related projects," says *Aviation Week*, "However the commercial aircraft divisions of the same companies are limited to research and development activities."
25. The air force has made much of its desire to "flow down" the benefits of MT programs to smaller subcontractors. But the first four subcontractor recipients of Techmod money ($1 million for F–16 components in 1982) could hardly be classified as struggling small businesses; they are Westinghouse ($1.1 billion in 1981 DoD sales), Sperry Rand ($928 million in 1981 DoD sales), Simmonds Precision ($139 million in overall 1982 sales) and Tracor ($371 million in overall 1982 sales).
26. Raymond J. Larsen, "Flexible Manufacturing: The Technology Comes of Age, *Iron Age* (September 7, 1981). Manufacturers of FMS equipment (Bendix and Kearney & Trecker are two of America's biggest) seem the very picture of the innovative vendors that the industrial preparedness lobby says DoD and the country desperately need. In fact, the companies abide by the free enterprise book so faithfully that they don't even share software secrets with one another. But real free enterprise causes real consternation, not applause, at the Pentagon. "One of the reasons that TACOM (the Army Tank Command) came to us was the fact that they felt U.S. vendors were too biased," says Brian Moriarity, head of Draper Labs' TACOM project. "We have seen instances where FMS suppliers have refused to incorporate machine tools built by others into their proposals. We even had one proposal involving machines from two builders that was rejected simultaneously by both because they didn't want to work with each other."

    "The Air Force can't force standardization by itself" a ManTech Officer told *Iron Age*, "but the potentially large users of FMS's can and that is part of the Air Force FMS effort, to get the big users to impose some reasonable form of standardization on the FMS market. Will U.S. systems builders refuse to cooperate to the point of refusing to supply hardware?

It's war right now, a cold war, but a war nevertheless. And we really don't know at this point in time how it will turn out."

Air Force ManTech Director Gary Denman scoffs at the war metaphor, saying, "If Kearney and Trecker has developed an automatic tool changer I don't think you'll see the services pressure them to share the information." But Denman acknowledges the information-sharing goals of MT investments and the undesirability of "a company being stuck with K&T forever if they buy K&T equipment that doesn't use Bendix software." Although he hasn't roused much enthusiasm, Denman says, "We are trying to put some pressure on the industry for [software] standardization. The alternative approach is for us to make some software that'll allow those other softwares to talk with one another. It's a problem but not a show-stopper." (Denman, author interview, October 1982).

27. James E. Ashton, "Integration of the Manufacturing System," in *U.S. Leadership in Manufacturing*, a symposium at the 18th Annual Meeting of the National Academy of Engineering, Washington, D.C., November 4, 1982, p. 107.

28. "Capability of the U.S. Defense Industrial Base," Hearings of the House Armed Services Committee and the Panel on Defense Industrial Base, 96th Congress, September–December 1980, pp. 430–431. These broader criticisms should be carefully distinguished from the more common and richly deserved caveats about the performance of automated equipment. These performance criticisms often convey the indulgent tone reserved for developments in their infancy rather than hardnosed questioning of the assumptions behind the particular piece of automation. Nevertheless, people who have to work with machines, such as the current crop of robots, know how exasperatingly huge the gap is between high-tech hype and reality. Joseph Engleberger, the president of Unimation, Inc., says,

> We have all these [performance] attributes, all of the programmability that I spoke of before and you say, "Gee, it must be easy, isn't it, to do assembly with robots?" Just so that you can see how difficult it still remains, I want you all to be able to do what I call "play the robot assembly game." You can do this at home; it is a low-budget game.
>
> First, rub petroleum jelly on your glasses and then tie one hand behind your back. If this particular assembly job requires two hands, get a friend to rub petroleum jelly on his glasses and tie a hand behind his back. Next, put a mitten on that one hand and then pick up chop sticks. You now have every attribute of an assembly robot today, and all you do is to assemble something according to detailed instructions.
>
> We still have some work ahead of us to beat the assembly game.

Engleberger, *U.S. Leadership in Manufacturing*, pp. 72–73.

29. Larsen, "Flexible Manufacturing."

30. "I've been struck by the fact that we're entering a second technological revolution in industry and agriculture that I don't really understand, to be frank," Tennessee Governor Lamar Alexander said in explanation of his

efforts to build a multimillion dollar Technology Corridor in Oak Ridge. "We're talking about complex, advanced ideas that go way beyond me and will still go way beyond me when I'm an old man" Terry McWilliams, "Technology Corridor Will Lure Industry," *Knoxville Journal*, June 18, 1982.

31. Lt. General Lawrence Skantze, address to Manufacturing Technology Advisory Group: Phoenix AZ, October 18, 1982.
32. Dr. Gary Denman, address to Manufacturing Technology Advisory Group: Phoenix, AZ, October 19, 1982.
33. "Integration Puts the Manufacturing Engineer in the Center of Things," *American Machinist* Special Report 746 (June 1982).

    People in the field disagree over the disciplinary pigeonholing of manufacturing. George Ansell, Dean of Engineering at Rensselear Polytechnic Institute, says, "It is unlikely that manufacturing will ever be a science." Allen Newell, a computer science professor at Carnegie-Mellon University, says,

    > If one believes that manufacturing is in the same position as chemistry or physics or even mechanical engineering, that simply is not true. There is no body of PhDs generating theories of manufacturing and devoting their lives to understanding the nature of manufacturing . . . one thing we must do is to see if we can find a way to create in the scientific world, on the campuses, the notion of manufacturing as a fit topic for scientific study.

    Ansell and Newell in *U.S. Leadership in Manufacturing*, pp. 134–135.
34. Dr. Jay Black, author interview, December 1982, and Leo Young, "The University Link," *Defense '83* (February 1983). Much of the current flurry of activity was prefigured by the National Defense Education Act of 1958 (amended in 1964) which charged colleges and universities with training a new generation of scientists and engineers for the 1960s aerospace boom.
35. Black, author interview.
36. Ibid.
37. Ibid.
38. B.H. Weil, "Research at Georgia Tech," *The Georgia Tech Engineer* (May 1947); John A. Griffin, "A Progress Report of Georgia Tech's Part In The War," submitted to President Blake R. van Leer, October 20, 1944, filed in the Georgia Tech library.
39. Annual Reports of the Georgia Tech Engineering Experiment Station, 1953 and 1956–57, available in the Georgia Tech library.
40. "Georgia Tech: Impetus to Economic Growth," a report to Georgia Tech National Alumni Association by Arthur D. Little, Inc., November 1963, available in the Georgia Tech library.
41. Gary Hector, "Atlanta Seeks High Technology," *Kingsport Times* (November 17, 1982).

42. Gary Hector, "Atlanta Seeks High Technology," *Kingsport Times*, November 17, 1982; William A. Schaffer and W. Carl Biven, *The Impact of Georgia Tech: Money, People, Ideas* (Atlanta: Georgia Institute of Technology, 1978), p. 153.
43. "Materials Handling Research Center Established at Georgia Tech," release from the Georgia Tech News Bureau, February 22, 1983.
44. Rudy Yobs, author interview, January 1983.
45. Wayne Hodges, "High Technology Overview" (Address to Fourth Annual Venture Capital Conference, Atlanta, Ge., November 1982). The ATDC has undertaken to incubate everything from stereo speakers to computer-aided design systems. But two of its earliest projects aimed squarely at military markets for microwave antennae and solid-state millimeter wave systems.
46. Yobs, author interview.
47. Donald J. Grace, *Annual Report of the Engineering Experiment Station*, 1981–1982, p. 21. Senior research engineer Jim Muller was EES' representative to CAM–I's electronics automation project in 1982.
48. With an Advisory Group headed up by the chairman of Auburn University's Electrical Engineering School, ECAM is headed into an implementation phase that should involve several large military electronics primes (Phase One participants included Boeing, General Electric, General Dynamics, Honeywell, IBM, RCA, and Magnavox). ECAM participant Alfred Robinson of Battelle Labs projects that though the shopfloor manifestations of ECAM may be five to ten years away, "when we do see these effects they will be revolutionary." Robinson, "Electronics Computer-Aided Manufacturing," in *ICAM Sixth Annual Industry Days Proceedings*, Air Force Wright Aeronautical Labs, New Orleans, January 1982, pp. 379–395.
49. Tze I. Chiang, *The Opportunity for Electronics Industries in Georgia* (Atlanta: Georgia Tech EES, April 1982).
50. Yobs, author interview.
51. Ibid.
52. Denman, author interview.
53. Jeremy Main, "Work Won't be the Same Again," *Fortune* (June 28, 1982). Curiously, automation and its impacts have not been widely incorporated into analyses of the product cycle, comparative advantage, or technology transfer. Economist Eileen Appelbaum addresses the subject in a paper prepared for a 1982 Office of Technology Assessment workshop on labor markets. Appelbaum projects that significant control of new manufacturing technologies by multinational corporations may accelerate the pattern of "developing countries specializing in the production of an increasing number of manufacturing products" and "productivity gains" occurring outside the United States. Appelbaum, "The Economics of Technical Pro-

gress: Labor Issues Arising from the Spread of Programmable Automation Technologies," in *Automation and the Workplace*, OTA Technical Memorandum (Washington, D.C.: 1983).

54. *Payoff '80: Executive Report, Manufacturing Technology Investment Strategy*, Air Force Systems Command (October 1980), p. 28. AFSC doesn't bother explaining how it arrived at the 44 percent figure; however, the Messerschmidt-Bolkow-Blohm fighter plane plant in Augsburg, West Germany, claims a 44 percent reduction in skilled machinists since the introduction of an FMS.
55. "America Rushes to High Tech for Growth," *Business Week* (March 28, 1983).
56. "ICAM Program Prospectus," p. 5.
57. John Mandl, "Competition for Skilled Manpower in the Machine Tool Industry" (Address to the 1981 Industrial College of the Armed Forces Mobilization Conference, Washington, D.C., June 4, 1981).
58. Harvard Business School analyst Shoshanah Zuboff quoted in *Main.*
59. Kolcum, "C-5A Wing Modification," p. 000.
60. Richard Wysk, author interview, December 1982, and Gordon Mayer, author interview, October 1982. To be fair, not everyone in the automation camp subscribes to the upgrading doctrine. ICAM's Gordon Mayer says,

> It's difficult for me to believe that any individual can sit on an assembly line, day in and day out and install five bolts on a car—the same five. And yet they do it. If I were given a choice I'd rather operate a damn robot than put the five bolts in the car. I'd rather operate the robot than drill holes manually down at General Dynamics because the drills are heavy. You're having a bad day or it's Monday morning and your quality can go down and you're knocked for that. With a robot you don't have any of that.
>
> On the other hand it is monotonous. Do you get a lot of satisfaction from it? I doubt it—the robot's doing the work, you're not. You're just watching him. You're helping him if he has some error. For a while it's interesting. There aren't many robots. It's neat; I work with a robot and you don't. How long does that last? I don't know, maybe a week, a month. In the end I don't think that being a robot operator versus being what you used to be is all that different.

61. Mayer, author interview.
62. No comprehensive analysis has been performed on the relationship between job content and productivity in newly automated defense plants. One reason for this analytical hole may be that economists often know as little as or less than the military about making things. There are a few exceptions, however. Eileen Appelbaum, "The Economics of Technical Progress," notes:

> Suppose that when a U.S. firm puts in a robotic installation, it replaces as many as six master machinists with one programmer plus three entry-level people whose function is to "load and unload," keeping things lined up for the robots. In Japan, on the other hand, let us suppose that master machinists are retrained and prepared

for positions as machinist/programmers. The machinist's job is transformed but not downgraded, and the machinist is ready for future changeovers. The initial cash flow advantage is gained by the American firms which have a less skilled and lower paid labor force. However, the man/machine configuration in U.S. firms is more permanent, less flexible. In the absence of skilled master machinists, the opportunities for learning by doing are severely curtailed. The Japanese in this example, because they retain their master machinists, need to build less into the machine, can design less immutable man/machine configurations, have enhanced opportunities for learning by doing, and have increased opportunities for continuous technological change. Longrun competitive advantage would rest with the Japanese. This example suggests that the substitution of less skilled workers for craft and highly skilled workers, as a means of holding down costs and increasing profits in the near future, may be myopic. In the longer period, it could have serious implications for international competitiveness and manufacturing employment.

63. Mayer, author interview.
64. *Human Factors Affecting ICAM Implementation*, Air Force Systems Command, Wright-Patterson AFB (September 1981). It is important to remember that the military's approach to human factors is not the only approach. For a dramatic contrast see *Strains & Sprains: A Worker's Guide to Job Design* (Detroit: United Auto Workers, 1982). The UAW (which calls human factors by its other name, ergonomics) undertakes to define human factors from the worker's point of view rather than as a product of management largesse.
65. "Larry Sturgill," author interview, January 1983.
66. See, for example, testimony of Brian L. Usilaner, Associate Director of GAO's National Productivity Group at "Automation in the Workplace: Barriers, Impacts on the Workforce, and the Federal Role," Hearings before the Labor Standards Subcommittee of the House Education and Labor Committee, 97th Congress, Washington, D.C., June 23, 1982. The gap between federal funding of military manufacturing technology programs and federal scrutiny of those programs continues to grow. In January 1984, a Congressional Budget study, *Federal Support of U.S. Business*, projected that Defense Department ManTech funding will grow faster than any of the federal government's myriad credit or direct expenditure programs between FYs 1984 and 1988. According to the CBO, military ManTech subsidies will jump 177 percent during that period.

    The OTA study is *Computerized Manufacturing Automation: Employment, Education and the Workplace* (Washington, D.C.: General Printing Office, 1984), pp. 309–319, 389.
67. Raymond J. Larsen, "Flexible Manufacturing: The Technology Comes of Age," *Iron Age*, (September 7, 1981).

10 CONCLUSIONS AND COUNTERCURRENTS

John Tirman

The overriding sentiment in the foregoing chapters is that defense procurement has a negative impact on commercial technology. Our view is buttressed by more than sentiment, however, and implicitly suggests a pattern that the relationship between the Pentagon and high technology assumes.

In describing this pattern, we give some coherence to what is often an analysis or opinion relying heavily on anecdote. It is common to hear nowadays of the military encroaching upon civilian institutions and projects—such as the space shuttle or the Jet Propulsion Laboratory—and to learn of instances of people within those settings seeking DoD funding in lieu of traditional, nonmilitary sources of support. The real-dollar reduction in nonmilitary research and development funding, coupled with the major surge in defense procurement, certainly leads one to the easy conclusion that private firms, public labs, and universities would naturally gravitate toward the Pentagon booty. Yet such intuitive reasoning, however valid, does not tell us much about what the effects of that trend are.

As was stated in the preface, any pattern described is merely suggestive. There is no "control group," no existing alternative to compare with the current state of affairs. High military spending has been a constant in the United States for four decades, and the Department of Defense has been the principal economic agent in the nation during that time. Some studies have contrasted America's flagging productivity growth, poor balance of trade, and other such indicators with West Germany and Japan, neither of which have a significant military budget. Such are spurious comparisons, because the elements of those nations' success are many—including good management-labor relations and national (civilian) planning—not simply the absence of a heavy military burden. So the conclusions forwarded here are derived from an empirical vacuum; it's not a perfect vacuum, however, because much can be deduced from the facts we know.

The pattern of effects indicated in the previous chapters, then, can be summarized as follows.

Defense Affects the Choices and Conduct of Research in the University, Industry, and National Laboratories. Because the Pentagon employs approximately one-third of the scientific and technical work force—with higher proportions in some crucial fields—and because the funding of research and development is more dependent on military sources than any other, defense is the dominant influence in the kinds of activities conducted in the nation's research centers. This is particularly evident in a time of declining research funds available for nonmilitary basic science. Increasing portions of the budget resources of national labs and Federal agencies (such as the Department of Energy and NASA) are being dedicated to military-related R&D, and declining portions of budgets are supporting activities in, for example, medical and energy research.

Even without the current trend in funding, however, the military has exerted a strong influence on research and development. This occurs principally through the Pentagon's ability to attract top scientific talent with high salaries and the promise of working on "sweet" technical problems. Major universities, many of which enjoy high levels of DoD support, may design curricula to satisfy particular needs of defense research. The funding mechanism both for individual researchers and institutions encourages a pursuit of science and engineering that contains defense applications, thus skewing, however subtly, the intent and conduct of a wide variety of research and development projects.

Defense Limits or Distorts the Traditional Prerogatives of Science. The commercial high technology industries are "science-based," reliant on technical innovation and new knowledge for growth. In a military research setting, the scientist's typical method of work is altered in two fundamental ways.

First, it is circumscribed severely by the military's insistence on secrecy. The lifeblood of science is publishing work, conference seminars, and informal exchanges around the world with one's peers. These normal channels of communication, criticism, and the refinement of data and methodology are closed to the military scientist. Moreover, the secrecy enforced within a specific project—the limit on the "need to know" the purpose of research or even the next stage of application—distorts the natural environment and scope of the scientist. The narrowness thus engendered can be a crippling constriction of a scientist's development and maintenance of skills.

Second, the "need to know" syndrome, coupled with general secrecy and the nature of military contracting, can create overspecialization. A physicist who has been designing instruments for radar signature analysis for a decade may have little to offer a commercial company. Such narrow development of knowledge has especially pernicious effects when the scientist leaves a military research setting—voluntarily or not—and seeks a job in a commercial firm or in pure research. In any case, it creates a large army of scientists who are suitable only for military-related work.

Defense Often Directs the Course of Technological Development. In addition to the haphazard influence on the types of research pursued in science centers, as explained above, the Pentagon can dominant a entire industry and the technology it nurtures. Commercial nuclear power is the classic example. Had the federal government, during the 1950s and 1960s, pursued several different reactor designs and cautiously considered the ways that nuclear energy could be integrated with the electricity system, the result may have been more fruitful than it has been. Even if left entirely to a free market, a better outcome may have been in the offing. Instead, the design fostered by the U.S. Navy was heavily promoted by the Atomic Energy Commission, a design flawed in many respects. The consequence, as we have seen in the 1980s, is a wholesale economic disaster compared with the widely held expectations for the technology.

Some analysts suggest that a similar corrosion has been at work in aviation, though the evidence is ambiguous. Aerospace has been consciously separated into a civilian and military effort in the federal government; a complaint sometimes articulated is that there has not been enough technology sharing between the two. Fortunately, commercial and military aerospace functions can be quite alike—cartography and reconnaissance, for instance. But the importance of outer space to the military has been growing in recent years, and the temptation for the Pentagon to dominate aerospace science grows apace. President Reagan's initiative for a space-based ballistic missile defense portends just such a dominance. The current plan, beginning with the fiscal year 1985 budget, is to spend $26 billion for research and development of space weapons during the last half of the 1980s.[1] That budget commitment represents a hefty rise for such activities.

How does that threaten "dominance" of aerospace? The first and most important reason is the drain of technical talent to the weapons effort. The advances in knowledge in this field will be purposefully military in nature and the knowledge may or may not have useful civilian applications, yet such applications may be so shaped as a result. Federal resources available for commercial activities may suffer as well. The private, commercial enterprises in space—materials processing, telecommunications, and so forth—may also be affected as a result of space becoming a hostile terrain.[2] In addition to the ballistic missile defense program, the United States and the Soviet Union are engaged in an antisatellite weapons rivalry; the U.S. Secretary of Defense has spoken of "hardening" satellites against attack. Space factories may be affected, or their missions retarded by the lawlessness of outer space. NASA's space shuttle is taking on more military functions as well. All of those factors may blunt the peaceful and profitable exploitation of space, and drive technical development steadily toward military purposes.

One can never predict with confidence, of course, but the elements of the military's dominance of aerospace are apparent. They differ from the way that the Pentagon influenced nuclear energy or aviation, but DoD's pivotal role in the industry is no less important.

Defense Adversely Affects the Small, Innovative Firms in High Technology. It is no secret that much of the innovation in science-based industries is derived from small companies. The boom in electronics and the visible advances in biotechnology seems to bear this out nicely. Defense procurement presents special dangers to such firms, whether or not they are involved in DoD work.

The principal threat is, again, the drain on technical talent. New firms especially cannot compete with salaries paid for defense work.

The recruiting network of defense contractors is broad and established. Once a scientist is in defense work, it may be difficult to get out.

For the small firm attracted by defense contracts, the dangers are palpable. Defense contracting procedures place an inordinate burden on small firms; the staff time and paperwork alone is substantial, certainly far greater than what is required for commercial work. Meeting technological specifications can be stressful. The cost-plus mode of contracting may create, over time, a laxness in personnel and procedures that will place the firm at a disadvantage when engaged in a purely private—and competitive—atmosphere. Strictures of military secrecy may engender operations that are not prudent for commercial work, including the aforementioned limitations on scientists' skills. And, not least, the boom-and-bust cycles of military contracting can literally bankrupt a small firm.

The last factor persuaded many high technology companies, large and small, to forgo the lure of military dollars following the last "bust" in the early 1970s. But the attraction is strong nonetheless, particularly when the economy is slack and military contracts are flowing, as occurred soon after Mr. Reagan took office.

Defense Specifications for High Tech Products May Create "Overdevelopment." One of the clearest and most profund aspects of the procurement process is that the technology used in military applications, even if it is largely identical to a commercial technology, requires technical specifications of far greater precision or of a wholly different nature than do commercial technologies. That is, even a mundane product may have to be "ruggedized," made stronger or more resilient to withstand extremes of temperature, motion, radioactivity, and so forth. Performance standards in general, including very sophisticated techniques, are much more extensive for military technology.

The trouble with such requirements is not merely the limited applicability to commercial enterprise, but the demand for a significant scientific and engineering commitment. (And there are those who seriously question the military utility of such sophistication.) The specifications mania is not restricted to this or that product or weapon, either: the larger fear is that it can afflict entire industries. Thus, it is held that American aviation is a prime case of "overdevelopment," an industry that has been driven to a technical excellence for (perhaps spurious) military reasons, but has not been capable of employing this level of sophistication gainfully to its commercial mission. It is conceivable that a parallel process is underway in electronics, signalled by the Defense Department's determination to develop the VHSIC (very high-speed integrated circuit).

Defense Has Enormous Effects on Economic Conditions Generally—Which Affect Growth Industries—Yet Has No Corresponding "Industrial Policy" to Compensate for Such Effects. This is the "macro" argument that is so common to the national debate about defense spending. There are several components of this view: (1) Pentagon spending, because it is not "productive," stimulates inflation by pumping wages into the economy without creating new products; (2) defense spending creates unemployment, because it is capital-intensive; (3) the political nature of defense spending has lead a shift of investment to the south and southwest, hastening the decline of the northeast and midwest and wasting those areas' "standing capital" of factories, homes, and schools; (4) the sheer size of defense budgets over time have sapped America's economic vitality and leads inevitably to tradeoffs in federal spending commitments, some of which (like education) have very direct impacts on science-based industries. There are many variants of these arguments as well.

A newer perspective is beginning to emerge that regards the Pentagon as the major economic actor in the U.S. economy. It sees its procurement, construction, personnel, and technological development programs as an inadvertent "industrial policy." The Department of Defense is engaged in virtually every aspect of the economy. It spends more money—by far—than any other single institution. It affects regional development through its procurement and deployment plans. It affects technological development in many ways. It affects the nature of labor markets, employment, and "deskilling" through its vigorous pursuit of automated manufacturing technologies, among other things. It affects the balance of trade and the exploitation of natural resources. There is very little that the Pentagon does not touch.

Yet despite this awesome influence, the Defense Department does not approach its missions with its own economic muscle in mind—except the political leverage employed routinely. Retraining workers, compensating for regional shifts of capital, underwriting the education of scientific personnel, and comprehensive planning for other macroeconomic effects is in scarce evidence. What thinking does go into these matters appears to be ad hoc, uncoordinated, ineffective, or simply too small in relation to the Pentagon's real effects. Political planning—spreading the goodies to gain congressional support—appears to receive far more attention in the Pentagon than economic consequences do.

The foregoing, then, is a possible "paradigm" for the impact of defense procurement on advanced technology. It is altogether a rather convincing case, but it is far from airtight. One shortcoming of the pattern presented here is that it cannot be decisively proved. Of course, those complaining on such grounds are equally at a loss to prove the opposite case. But at least three major objections can be added to the general (and unresolvable) criticism regarding empirical evidence.

The first is that defense spending for technology creates many commercial applications that would not necessarily arise in the absence of DoD involvement. This is the "spinoff" argument, and it is engaged frequently throughout the chapters of this book. It is an assertion that cannot confidently be put to rest.

Without a doubt, military research and development do spin off commercial products, and in impressive quantities and quality. This phenomenon is sure to continue. The doubts about this process are raised in three contentions. The first is that many of the products made available from spinoffs would have been developed through private means anyway; the Defense Department perhaps hastens the development of such gadgets (and may distort development in some instances), but the supposition that military funds are wholly responsible for the creation of such products is spurious. The second point holds that the major benefits of the spinoff phenomenon have already accrued; DoD money helps most when a technology is in its infancy. As technologies and industries mature—and, indeed, as they are "overdeveloped"—the spinoffs are less and less frequent. The third assertion is that *any* public agency spending the kind of money the Pentagon spends on technology is bound to create some commercial by-products. If spinoffs are wanted, there are better public policies and institutions available to achieve that objective—something along the lines of Japan's MITI, for instance.

The last of these retorts to the spinoff argument, however, has a rejoinder: in a free-market economy, where heavy government involvement in the promotion of industrial technology is frowned upon, the possibilities of having a MITI-style federal agency are practically nil. Thus, the Defense Department actually serves a useful, if unintended, role in economic development and technological advances.

The third possible flaw in the defense critics' position is that the military's emphasis on high technology and its procurement practices are simply necessary for national security. This, of course, is the argument most often relied upon by civilian and military officials in their justification for high military budgets and the growing emphasis on high-tech weaponry. We will return to this consideration in the last section of this chapter.

THE DARPA EXCEPTION

While the lion's share of military R&D expenditures contributes little to economic development, and may even detract from it, there are exceptions to every rule. We would be remiss if we did not highlight one outstanding exception in the Department of Defense. In the major research universities, one of the most respected and sought after government research agencies is DARPA—the Defense Advanced Research Project Agency.

DARPA, also referred to as ARPA—Advanced Research Projects Agency—was born in the wake of Sputnik. Dismayed at being caught flat-footed by the Russians, President Eisenhower asked how such a strategically important advance could fall through the cracks. The answer was partly organizational. No agency or service was accountable to such revolutionary long-term research projects. Thus, DARPA was established to cut across the traditional jurisdictions of the armed services and to maintain a technological vigil to assure that the technology available for national security was second to none.

Initially DARPA spearheaded the space race. Shortly thereafter the National Aeronautic and Space Administration was established, and selected DARPA personnel were transferred to the new agency, including the Saturn-booster program. Today, the work of DARPA, with a budget in FY 1984 of approximately $877 million and a staff of 150, is divided among five offices. Of these, the Information Processing Technologies Office (IPTO) is particularly relevant to our discussion of the role of government in innovation and international competitiveness.

This small IPTO group, presently numbering twelve professionals with a 1984 budget of $120 million, has had a disproportionate impact on the progress of computer science and information processing over the past twenty years. For example, through IPTO's funding of ARPANET, a packet-switching technology was developed which now serves as a cornerstone of modern data-communications networks to link computers in a wide range of commercial and industrial, as well as military, applications. Time-sharing technology, developed by Project MAC at M.I.T. and others, was largely funded by IPTO. The development of the Illiac IV computer was funded by IPTO, and while only one copy of the machine was ever built, few would deny that its underlying technology, parallel-processing architecture, and advances in memory laid the groundwork for modern-day supercomputers. Perhaps most significant, IPTO has been the major factor in sponsoring the development of artificial intelligence which lies at the heart of the fifth generation computer efforts today.

The progress and innovation sparked by DARPA is truly impressive, especially in light of its relatively modest budget. What is the model here? What has permitted a government agency such effectiveness, and can it be extended to other areas of government or industrial involvement? By far the most important element of DARPA's success has been the quality and leadership capabilities of its staff. IPTO, for example, has been able to attract a long line of world-class talent. The list of past directors includes such scientists as J.C.R. Licklider, I.E. Sutherland, L.G. Roberts, and now R.E. Kahn. Each has earned international recognition for significant technical contributions, and all are held in highest respect by their peers in industry and universities. Bob Kahn, a former professor at M.I.T., masterminded the ARPANET development and steadfastly fostered packet-switching technology over the past fifteen years, most of which were at DARPA. He has been associated with IPTO for the past eleven years, almost twice as long as any of his predecessors. It is no exaggeration to say that Kahn has played a truly significant national role in planning, funding, and directing a large portion of the United States research effort in artificial intelligence, mostly in universities, but selectively in U.S. industry as well.

Another factor in DARPA's success is that its relatively small staff is largely unencumbered by bureaucratic processes. DARPA's philosophy is to select competent, proven staff professionals and then to delegate authority to make significant and far-reaching decisions. Innovation is ultimately a risky business, and any effort to mitigate this risk through collective committee action compromises the cutting edge of technological advance. To quote Kahn, "Our approach is to identify the most talented researchers to work on critical scientific and technological problems of interest to defense. In the final analysis, our investments are in the key people working on first rate ideas."

A unique aspect of DARPA's strategy is its visionary focus on the potential end results of research. For example, putting a man on the moon or establishing a worldwide network for secure computer communications are the kinds of mission statements that inspire imagination, demand innovation, and pull together a critical mass of related activities. The output of generic research that results from these mind-stretching goals is far greater than the limited results of military R&D aimed at designing and building weapons systems. A great deal of DARPA's innovative strength is derived from the fact that it is not directly accountable to the armed services as end users, nor are they beholden to a rigid schedule that would compromise research goals. DARPA is also permitted to fail on some of its projects, which allows it to take the high risks others might steadfastly avoid.

While less than 10 percent of the overall DARPA budget is earmarked for university research, a unique aspect of IPTO is that more than half of its budget is invested in university research. The bias of DARPA toward quality over quantity is reflected in its high concentration of funding among a handful of major research universities. M.I.T., Carnegie-Mellon, Stanford, and more recently, Berkeley, the leading universities in computer science, owe much to DARPA for their prominence in this field. Not only has DARPA been a consistent and reliable source of long-term research funding for computer science, but it has provided a sense of direction and discipline to knit together into a cohesive pattern the overall effort on various campuses.

By concentrating its funding in a few centers of excellence, both in universities and in industry, DARPA has managed to generate a sufficient mass of resources to take on complex and challenging research projects. There is no question that they have been successful in raising America's research capability in computer science head and shoulders above international competitors. A major goal in concentrating on leading research universities has been not only the development of innovative ideas, but also the development of the needed human resources—trained scientific personnel and leadership talent. In an emerging new field like artificial intelligence, it takes a decade or more of consistent funding to build up a network of people who are conversant with a new generation of knowledge. Thanks largely to DARPA, America is now ready to move in bold new directions in information processing.

Twenty-five years ago, DARPA led the charge to restore our technical leadership in the space race against the Soviet Union. The pieces were there but they had to be pulled together and focussed on a mission. Now we are faced with another race—this time with the Japanese—for leadership in information processing. It took Sputnik to jar us into action then. It took Japan's Fifth Generation Computer Project to awaken our competitive spirit now. Again DARPA has seized the initiative. Its strategic computing program with five-year funding of $600 million was recently submitted to the Congress. It lays out a multiyear research and development program to assure U.S. superiority in information processing. Funding for the first year was set by Congress at $50 million, with $95 million proposed by DARPA in FY 1985 and $150 million the following year. The basis for this focussed effort is national security, but the potential for strengthening our industrial base is also immense.

By default, we are dependent on our military machine to cover our economic flanks. Will it work this time? One issue of concern is whether DARPA can continue to attract and retain a staff of world-class calibre when salary differentials between government pay scales and private industry continue to diverge. Another concern is whether DARPA will be able to maintain its highly efficient, nonbureaucratic process for a program that will reach funding levels of $150 million per year by 1986. Further, will the strategic computing program meet military needs while at the same time providing transfers of the generic technology to commercial and industrial applications? And at a time when top computer talent is so thin, will this program serve only to rob Peter to pay Paul? For example, if the budgetary tradeoffs in DARPA to accommodate the strategic computing program result in a reduction in funding for a buildup of computer science competence, the underpinnings of the program itself could be jeopardized and the long-term investment in university research severely eroded. It would be especially valuable if the funding were broadened to include the so-called second tier universities, given that M.I.T., Carnegie-Mellon, and Stanford are saturated and unable or unwilling to expand further in this field.

Notwithstanding these concerns, on balance it is comforting to see DARPA take the lead in providing a coordinated response to the Japanese challenge. At the same time, it is disturbing to know that the federal government is addressing our economic security as if it were a sideshow in our quest for military superiority. Why does the U.S. response to international challenges have to be couched in defense terms? The United States needs to launch a major industrial consortium for research, one that can draw on the best our universities have to offer, in conjunction with corporate partners. It should be funded not just by the high technology companies in computers and communications, but also by the benefactors of the new technologies. Auto and steel companies and the banking and insurance institutions should be willing to rally around a new effort to put in place a program of truly national significance in the information-processing field. An effort like this could be leveraged and matched with government funds, but the initial direction needs to be set by the private sector. This would represent a giant step toward improving our innovative capacity and directing our dollars toward civilian research and development. These steps, combined with civilian education and training, can create a population prepared for an economy of the future that is abundant in technology—for civilian uses rather than military ones.

James Botkin and Dan Dimancescu

BIOTECHNOLOGY AND THE MILITARY

Since World War II, three major high technologies have emerged. Aviation and aerospace have achieved remarkable levels of sophistication, though it is probable that commercial aviation has reached a plateau. The failure of the French and British supersonic transport, a technology started as a military project and subsequently promoted in the 1960s as a commercial boon (fortunately, the Congress refused President Nixon's desire to promote the SST), may be the last major innovation in aviation's history. Aerospace still has many frontiers to explore, including space factories. The second major technology of the postwar era, nuclear power, is at a dead end commercially in the United States. Its prospects abroad are less distinct, and it is conceivable that some variant of the technology—including fusion energy—may be employed in the 21st century. The third prominent innovation has been microelectronics, whose potential has only begun to blossom in the last decade.

Many analysts are now certain that biotechnology represents the fourth technological revolution of our time. It is worth considering, briefly, how this promising new "revolution" may be affected by the military, using the paradigm we have suggested in this book.

What is now commonly meant by "biotechnology" is recombinant DNA and cell fusion, and large-scale production through "bioprocess technology." These methods are distinguished from the ancient practices of animal husbandry, selective breeding in forestry, and the like. The new biotech has received its greatest impetus from medical research and the pharmaceutical industry, but agricultural, chemical, and even electronics industries are now playing a significant role in research and development.

The structure of the industry resembles that of electronics: small, innovative firms—dozens of which spring up annually—compete with older corporations that have been manufacturing and marketing product lines of similar (and more primitive) functions. The number of old versus new companies engaged in biotechnology applications appears to be roughly divided in half, though the older firms tend to dominate marketing, the newer, entrepreneurial firms seem to be the research-oriented innovators. In 1983, approximately $1 billion was spent to commercialize new biotech products in the United States by 219 firms.[3] One analysis suggests that 100 new firms will be set up in 1984 to develop new products at an investment cost of $2.5 billion, and that sales before the end of the century will top $15 billion.[4]

The funding role of the federal government has been largely in basic research ($511 million in fiscal year 1983), and is quite small in generic applied research ($6.4 million in FY83); the latter is probably less than 0.5 percent of the private sector's investment in R&D.[5] Aside from controversies over the safety of rDNA research and the role of the National Institutes of Health in setting guidelines for such research, the principal area of worry in biotech R&D seems to center on the university-industry relationship and its effect on science. Ironically, and perhaps revealingly, the concern about this relationship mirrors some of those of the military-high tech interaction. One report, for example, notes a "drop in communication between scientists, as university researchers align themselves with competing companies," and a decline in independent scientists who can advise policymakers.[6]

It is difficult to say precisely what the military's interest is in biotechnology, but it is interested. Military uses of recombinant DNA—or rDNA—in particular (which seems to be the most applicable to warfare of all biotechnology processes) is banned by the Biological Weapons Convention of 1972. Specifically, the treaty—to which the United States and the Soviet Union both are parties—prohibits the stockpiling and production of harmful biological agents, and subsequent definitions of such agents have clearly included recombinant DNA techniques.[7] Like most treaties, however, the 1972 convention does not prohibit *research*, and it is within this area that much military-related activity can be and is being pursued.

The research aspects of DoD's interest in biotechnology involve a paradox. The U.S. Government has stated that it interprets the convention for "defensive purposes." That may include, in the words of Henry Kissinger, then National Security Advisor to President Nixon, "research into those offensive aspects of bacteriological/biological agents necessary to determine what defensive measures are required."[8] In other words, immunization from (and deterrence of) an enemy's biological weapons necessitates knowing what those weapons are.

In 1980–83, the Pentagon undertook at least fifteen unclassified biotechnology projects, nine of which have been contracted outside military institutions, and all of which fall roughly into the categories of vaccine or "detoxification" research.[9] Three of the nine contracts were at nonuniversity institutions–one a firm in Minnesota, another at the Salk Institute in San Diego, and the third at the Weizman Institute in Israel.[10] The DoD funding has increased 54 percent in that period, reaching $100 million in 1983.[11] Some observers suggest that the Defense Department may be accelerating its rDNA and cell fusion programs in response to the fears of biological and chemical warfare being waged in southeast Asia and in central Asia. Although the controversy over "yellow rain" in Cambodia (the allegation that the Soviet-backed Vietnamese are using toxins), and chemical weapons use in the Iran–Iraq war are not substantiated, the mere spectre of such conflicts may drive the Pentagon to redouble its efforts.

In any case, unclassified evidence certainly points toward a fairly large and growing DoD biotechnology program, one that straddles definitions of offensive and defensive purpose.[12] We can then speculate, using the pattern outlined earlier in this chapter, as to the potential effect of a large DoD program on the commercial biotechnology industry.

The defense biotechnology budget does not appear to be so large that it will dramatically affect scientists' choices of research. On present trends, the industry is growing faster than DoD's biotech programs, though, by current standards–and compared with other high tech fields–the DoD share of total U.S. R&D in biotechnology is rather substantial. More worrisome, perhaps, is the decline of other relevant federal funds. Pentagon grants to universities for biological research increased 24 percent in real terms from 1980 to 1982, while other federal research grants dropped in the same period, which certainly suggests that many scientists would be tempted to apply for military money.[13]

Nor does it appear that DoD can now direct the course of biotechnology development. The field is now quite fluid, in a state of exploration and discovery, and such fluidity would seem to militate against dominance by any single institution. At the same time, however, biotechnology is also in its infancy, a small industry in large measure made up of small firms. When a "shake-out" occurs–when natural market forces combine to favor the commercially promising and to starve others of capital–many companies could be driven to the Pentagon to stay afloat. Thus, one can imagine a growing sector of military-industrial firms engaged in biotechnology research and development, and such a sector–by paying higher salaries, imposing secrecy, and the like—could come to influence greatly the industry as a whole. As it now stands, however, the Defense Department is contracting little of its biotech work to private institutions. For the military to dominate the industry, therefore, its commitment to biotech R&D would have to enlarge significantly *and* its present method of conducting that research would have to expand to include private firms. A large civilian sector that was producing toxins would also have to overcome the resistance to such research that has already appeared in many communities throughout the United States. The regulatory constraints alone could discourage the development of such a sector.[14]

Perhaps a more visible danger entails the "overdevelopment" argument. This includes two elements: that technology transfer between military and civilian sectors is insufficient (or diminishing), and that in certain industries, such as aviation, the military has driven development too far, making it too specialized for commercial application. Again, the prospects for this phenomenon recurring in biotechnology are not entirely clear. But the present program of DoD research suggests that little transfer may be forthcoming. The military is apparently concentrating on research to create vaccines against toxic and pathological agents that do not present a typical health hazard to a peacetime population. Although much commercial biotech is also in medical applications, other applications are beginning to gain larger shares of commercial R&D. It is conceivable, of course, that vaccine research needn't be directly applicable to commercial possibilities to create spin-offs in procedures, research design, production, and the like, but the apparent disjuncture between the research goals of the military and the disparate goals within the industry promises little technology sharing.

The additional elements of our high-tech-military pattern can apply equally to biotechnology. In particular, the effects of military contracting on small firms and on the scientists engaged in military-related R&D is always worrisome.

Ultimately, the effects of the Defense Department on biotech would seem to be circumscribed by the limited potential of the technology for military purposes. Whereas aviation and electronics are now the very foundation of the military enterprise, such does not seem to be the likely fate of cell fusion and rDNA. There is a comprehensive treaty in place to prevent such utilization, and it is clearly not in the interests of the superpowers to abrogate that accord. It is also possible, however, that new applications of biotechnology will create fresh military demands. What is now called "bioelectronics"—the development biosensors and biochips—is now an imaginable use of biotechnology not directly related to bioprocesses being employed as weapons per se. One can foresee a strong DoD attraction to these possibilities.

One element of the general course of high technology development is already showing itself with great clarity: competition from Japan. "The Japanese consider biotechnology to be the last major technological revolution of this century," an Office of Technology Assessment report warns, "and the commercialization of biotechnology is accelerating over a broad range of industries, many of which have extensive bioprocessing experience."[15]

As with the other high technologies in their infancy, the outlines of biotechnology's prospects are vague, and so the military's role in their future development are equally vague. But many of the elements in our paradigm are present, and thus the potential for damage to biotech's commercial vitality are evident.

SOLUTIONS

It was not the purpose of this book to present detailed solutions to the perceived problems we have outlined, but a few suggestions are worth mentioning. Our brief treatment of alternatives to the current state of affairs falls into two categories—the large and the small.

The first regards the nature of U.S. security needs. It is evident that the basis of the military's incursion into the world of high technology is motivated by its responsibility for national defense. Yet the actual dimensions of that responsibility and the security needs underlying the Pentagon's actions are a constant source of a very heated and complex debate. It is not possible to engage that debate here, but a few elements of it are particularly relevant and can indicate what I mean by a "solution."

A substantial and growing amount of the high tech used in the armed services is dedicated to enlarging the capability of weapons, particularly nuclear weapons, and the support systems—satellites, radar, C^3I, and the like—which enable weapons use. On the surface, a continuous trend toward increasing military capability appears compatible with the Defense Department's mission to increase security. More accurate missiles, more certain guidance, and greater speeds, range, and variety of attack modes all seem to fulfill the Pentagon's responsibility to bolster U.S. military strength, with the ultimate aim of deterring aggression from real and potential adversaries, the Soviet Union chief among them.

This apparent wisdom, which equates high capability with security, is now under attack. The doubt takes two forms. First is the persuasive evidence marshalled by analysts such as Mary Kaldor and James Fallows, indicating that advanced technology may actually weaken the performance of weapon systems and thereby diminish the readiness of U.S. forces. The second line of this argument is less provable but potentially more serious: the advanced technology of nuclear weapons creates, paradoxically, new vulnerability alongside its awesome capability.

The unambiguous trend in American nuclear forces is toward multiple-warhead missiles that have prompt, hard-target kill capability. An ICBM may have as many as ten warheads each, every one of which can hit a Soviet "hard target"—a missile silo or command post, for example—with great accuracy and in a very short time. That gives these missiles a first-strike capability: they can knock out the missile silos of the enemy, disarming and perhaps defeating it in a single, horrible stroke. The MX and Trident II missiles will have such capacity, and the Soviets are reciprocating.

At the same time, however, it is quite difficult to deploy such weapons in "invulnerable" basing modes, as the controversy over siting the MX in a racetrack or densepack configuration attests. The MX will be put in its predecessor's silos, which, it is widely agreed, are vulnerable to Russian ICBM attack. The result: an extremely powerful weapon, the MX missile, representing an immense threat to the Soviet Union's land-based missile force (which is 75 percent of the USSR's total nuclear arsenal), will be deployed in vulnerable U.S. silos. This combination—a highly capable weapon that is vulnerable to attack—is what arms analysts refer to as "destabilizing." The MX, because of its advanced technology, is a lucrative target for a Soviet first strike during a crisis that Soviet leaders regard as unmanageable. The U.S. sea-based deterrent, which will depend on twenty Trident submarines carrying up to 336 Trident II warheads on each boat, could be subject to identical problems if antisubmarine warfare techniques advance.[16]

The sum total of this trend is, as both Frank Barnaby and Warren Davis explained in their chapters, the potential for *less* security rather than more. When one includes such highly capable weapons as the Pershing II missile being deployed in Europe, Stealth bombers, and antisatellite weapons, it is not difficult to consider this danger as imminent. If space-based ballistic missile defense is also pursued, the United States (and, quite possibly, the USSR) will have moved far away from the relative stability of deterrence to a provocative arms race of uncertain prospects.

Thus, the easy assumption that more high technology integrated with military forces equals greater security is not only fatuous, but dangerous. Certain uses of computers, lasers, imaging, advanced avionics, and the like can increase security, of course. The space-based reconnaissance that enables arms control verification, early warning of attack, and similar functions may enhance stability and security. One can argue, too, that improved conventional weapons may improve deterrence and lessen the likelihood of using nuclear weapons, particularly along Europe's Central Front. But the notion that more accurate and prompt nuclear missiles, or airplanes that can escape radar, actually enhance U.S. security is not convincing. Yet this is the ultimate rationale for the Defense Department's huge demands on the high technology industries.

The "solution" to this involves tasks that go well beyond the scope of this book to address much larger questions of war and peace. A solution of that magnitude entails a broad imperative for the United States to reaffirm its commitment to deterrence alone, while pursuing arms control that can reduce significantly the requirements for deterrence. Such an arms control agenda, in the immediate sense, includes a comprehensive nuclear test ban, a ban on antisatellite weapons, deep cuts in strategic and intermediate-range nuclear forces, a negotiated shift in the U.S. force structure to a single-warhead missile deterrent (which would require the cancellation of the MX and Trident II), and stringent measures to curb nuclear weapons proliferation.

That is the large solution, which is large indeed. More in keeping with the project of this book are "small" solutions aimed at mitigating or preventing the inimical economic impacts of military spending.

It should be noted that there are large areas that this book has neglected: secondary education and military conversion, in particular. We have addressed various aspects of higher education throughout the book, but have no cohesive analysis, and there is little mention of secondary education at all. Needless to say, in speculating on the future of science-based industries, education is a key consideration. The role of training of all types in the high technology industries—and the military's influence on education—is a vast subject of equally vast significance. It deserves its own book.

Another area related to solutions is what is commonly called military conversion: the transformation of military facilities, firms, industries, and so forth, to commercial enterprise. Seymour Melman and Marian Edelman, among others, have outlined the possibilities for conversion.[17] Although somewhat outside the scope of this study—it requires a prior commitment to lower military spending, and needn't involve high tech or procurement per se—conversion obviously has great promise and relevance to the ultimate solutions of the problems we have raised.

The foregoing chapters of this book also tacitly suggest measures that deserve detailed consideration. They would include: (1) DoD gear its use of high tech to reliability and "off-the-shelf" products rather than the relentless development of technologies with high-performance specifications and dubious military value; (2) the federal government consciously account for DoD's impact on regional development, labor, education, and the like, and pursue policies to lessen the impact of DoD's unacknowledged "industrial policy" as a matter of national security; (3) procurement procedures and schedules be designed with the health of high tech sectors and trade as a priority; (4) the secrecy imposed on science be reviewed periodically by independent scientists and Congress; (5) consideration be given to a restructuring of the defense industry as a whole, to increase innovation, lessen political pressures on members of Congress and localities, and revamp the employment and contracting process to minimize the overspecialization impact on small firms and the technically skilled.

Such suggestions are only small by comparison with the larger arms control agenda outlined above. Some of these measures—and many others that are necessary to lessen the economic effects of the Pentagon—could be instituted. Others, such as the reconstruction of the "military-industrial complex," may be inaccessible to the ways and means of the American political system. A third avenue—demanding that our trading competitors and military allies, West Germany and Japan in particular, pick up more of the cost of Western security—has a kind of crude appeal, but is not a route to resolving the intrinsic shortcomings of the U.S. military procurement leviathan. In any case, we are only beginning to understand the problem; the realization of major solutions do not appear promising at this juncture.

But realizing the problem would be a very large step. Instead of the fiesty denials of the Pentagon's inimical economic impact that routinely emanate from the Secretary of Defense's office, we can hope for a cold, hard analysis of the current and future effects of defense procurement on advanced technology. The more that is learned about technological innovation and its relationship to military needs, the more we are likely to seek reform.

NOTES TO CHAPTER 10

1. Charles Mohr, "Study Assails Idea of Missile Defense," *New York Times* (March 22, 1984), p. 11. A fuller analysis of the costs of ballistic missile defense can be found in Kurt Gottfried et alia, *Space-Based Missile Defense* (Cambridge, Ma.: Union of Concerned Scientists, 1984).
2. See William Rosenau, "Approaching High Noon in Space: Satellite Weaponry May Crowd Out Business Development," *Los Angeles Times* (March 15, 1984), p. 7.
3. Office of Technology Assessment, *Commercial Biotechnology: An International Analysis*, Summary (Washington, D.C.: Congress of the United States), p. 6. Hereafter, OTA.
4. "Biotech Comes of Age," *Business Week* (January 23, 1984), p. 84.
5. OTA, pp. 17-19.
6. Edward Dolnick, "How Will Industry Money Affect Research," *Boston Globe* (November 7, 1983), p. 43.
7. *Arms Control and Disarmament Agreements: Texts and History of Negotiations* (Washington, D.C.: U.S. Arms Control and Disarmament Agency, June 1977), pp. 113-123.
8. Cited in Susan Wright and Robert L. Sinsheimer, "Recombinant DNA and Biological Warfare," *Bulletin of Atomic Scientsits* 39: 9 (November 1983), p. 24.
9. Ibid., pp. 22-23.
10. Ibid.
11. Charles Piller, "Ultimate Vaccines: DNA—Key to Biological Warfare?" *The Nation* (December 10, 1983). The author cites the National Science Foundation as the source for funding figures. Piller also claims that he was given a DoD document listing thirty-seven rDNA projects being funded by the department, as opposed to the sixteen cited by Wright and Sinsheimer.

12. Wright and Sinsheimer, "Recombinant DNA and Biological Warfare," p. 22.
13. Ibid., p. 21–22.
14. Piller, "Ultimate Vaccines," p. 601.
15. OTA, p. 16.
16. "Intercontinental Weapons," Briefing Paper 4 (Cambridge, Ma.: Union of Concerned Scientists, August 1983), p. 3.
17. Suzanne Gordon and Dave McFadden, eds., *Economic Conversion: Revitalizing Americas Economy* (Cambridge, Mass.: Ballinger Publishing Company, forthcoming).

INDEX

ABOUT THE EDITOR

John Tirman is senior editor and head of the communications group of the Union of Concerned Scientists, Cambridge, Massachusetts. Tirman holds a Ph.D. in political science from Boston University, where he was also a Teaching Fellow of the College of Liberal Arts, 1972–75. He has worked as a political consultant, a reporter for *Time* magazine, and as senior policy analyst for the New England Regional Commission. His articles on energy, arms control, and science policy have appeared in *Technology Review*, the *New York Times*, the *Washington Post, The Nation*, and many other magazines.

LIST OF CONTRIBUTORS

Frank Barnaby is guest Professor of Peace Studies, Free University, Amsterdam, Consultant to the Stockholm International Peace Research Institute (SIPRI), and codirector of Just Defence. Between 1971 and 1981, he was Director of SIPRI, and prior to that was Executive Secretary of the Pugwash Conference on Science and World Affairs. Earlier, he was a research nuclear physicist at University College, London, and worked at the Atomic Weapons Research Establishment, Aldermaston. He is the author of *Man and Atom, The Nuclear Age*, and *Prospects for Peace*, and has written or edited several books and articles on disarmament and military technology.

James Botkin holds a doctorate from the Harvard Business School in computer-based systems. He is author of the Club of Rome report, *No Limits to Learning*, and is coauthor of *Global Stakes: The Future of High Technology in America.* He is cofounder of the Forum Humanum, an international network on future research, and former academic director of the Salzburg Seminar. Dr. Botkin is a consultant in the Technology & Strategy Group in Cambridge, Massachusetts.

Warren Davis is an award-winning physicist and consultant based in West Newton, Massachusetts. He earned his Ph.D. from the Massachusetts Institute of Technology, and for several years developed military technologies at the MIT Instrumentation Lab, MIT Lincoln Laboratory, and in private industry. In the early 1970s, he was consultant to the Max-Planck-Institut in West Germany, and in the early 1980s was staff physicist at the Smithsonian Astrophysical Observatory in Cambridge, Massachusetts. Dr. Davis was the cofounder and first president of the High Technology Professional for Peace.

Robert DeGrasse was Fellow with the Council on Economic Priorities, New York, and is the author of *Military Expansion, Economic Decline.* He holds an M.P.A. from Harvard's Kennedy School of Government.

Dan Dimancescu studied at Dartmouth, the Fletcher School of Law and Diplomacy, and the Harvard School of Business Administration. He edited *Rites of Way*, an analysis of changing U.S. transportation priorities, and is coauthor of *Global Stakes: The Future of High Technology in America.* He is a consultant and writer on high technology strategy and policy.

Lloyd J. Dumas is an associate professor of political economy at the University of Texas, Dallas. He is former member of the Committee on Science, Arms Control and National Security of the American Association for the Advancement of Science. Dr. Dumas has contributed to many books and magazines on technology and public policy.

Robert Reich is a member of the faculty of the Kennedy School of Government at Harvard University. He has been director of policy planning at the Federal Trade Commission (1976–81) and assistant to the U.S. Solicitor General (1974–76). He holds a law degree from Yale Law School, a masters in economics from Oxford University, England, and a B.A. from Dartmouth. He is coauthor of *Minding America's Business*, and is author of *The Next American Frontier.*

Tom Schlesinger staffs the Highlander Center's Defense Industry and Strategic Minerals Project in New Market, Tennessee. He was a consultant for the PBS *Frontline* documentary "Pentagon Inc.," and has contributed to several newspapers and magazines on issues of the military and regional development. He is the author of the Highlander report, *Our Own Worst Enemy: The Impact of Military Production on the Upper South.*

Gordon Thompson is Executive Director of the Institute for Resource and Security Studies, and is a consulting scientist to the Union of Concerned Scientists in Cambridge, Massachusetts. He holds a doctorate in mathematics from Oxford University, England. He is coauthor of *A Second Chance: New Hampshire's Electricity Future as a Model for the Nation*, and has written widely on technology, public policy, and arms control.

John Ullmann is professor of management at Hofstra University. He is an industrial engineer, specializing in the relationship between business, technology, and society. He has written extensively on innovation, industrial development, and the economic and political implications of the arms race. He is the author of texts on quantitative analysis and production management. Prof. Ullmann is also the secretary of SANE, a peace organization.